BIGFOOT

Encounters in New York & New England

Whitehall artist Eric Miner created this sketch of the creature seen in the Whitehall area and other areas based on various eyewitness reports.

BIGFOOT
Encounters in New York & New England

Documented Evidence, Stranger than Fiction

Robert E. Bartholomew, Ph.D.

Paul B. Bartholomew, B.S.

Edited by Christopher L. Murphy

ISBN-10: 0-88839-652-X
ISBN-13: 978-0-88839-652-5

Cataloging in Publication Data

Bartholomew, Robert E.
Bigfoot encounters in New York & New England : documented evidence, stranger than fiction / Robert E. Bartholomew, Paul B. Bartholomew ; edited by Christopher L. Murphy.

Includes bibliographical references and index.
ISBN 978-0-88839-652-5

1. Sasquatch—New York (State). 2. Sasquatch—New England. I. Bartholomew, Paul, 1964– II. Murphy, Christopher L. (Christopher Leo), 1941– III. Title.

QL89.2.S2B37 2008 001.944 C2008-902002-2

Printed in Indonesia — TK Printing

Editing: Christopher L. Murphy, Theresa Laviolette
Production: Christopher L. Murphy, Ingrid Luters
Cover design: Ingrid Luters
Front cover image: sketch by Eric Miner
Back cover photos: Robert and Paul Bartholomew

Published simultaneously in Canada and the United States by:

HANCOCK HOUSE PUBLISHERS LTD.
19313 Zero Avenue, Surrey, B.C. Canada V3S 9R9
(604) 538-1114 Fax (604) 538-2262

HANCOCK HOUSE PUBLISHERS
1431 Harrison Avenue, Blaine, WA U.S.A. 98230-5005
(604) 538-1114 Fax (604) 538-2262

***Website:* www.hancockhouse.com**
***Email:* sales@hancockhouse.com**

Dedication

This book would not have been possible without the support of our parents Emerson and Mary Bartholomew; the inspiration of the late Dr. Warren L. Cook; the long-time friendship of our colleague Bruce Hallenbeck; the courage of his late grandmother Martha Hallenbeck; the pioneering research of Bill Brann and Ron Lewis; the proofreading skills of Michael Pluta; and the perfectionist guidance of Christopher L. Murphy.

Contents

Part 3 Evaluating the Evidence

Part 4 Case Catalog: Bigfoot-Related Incidents

Acknowledgments

We wish to thank the following researchers: Rick Berry, William Chapleau, Janet and Colin Bord, Ray Cecot, Loren Coleman, Paul Cropper, Red Elk, John Green, Tony Healy, Ben Jamison, Yasushi Kojo, Steve Kulls, Barbara Malloy, Gary Mangiacopra, Matt Moneymaker, Brad and Sherry Steiger, Warren Thompson, the late Ed Warren, Lorraine Warren, Cheryl Wicks, and Joe Zarzynski.

Thank you to the following people: Kaye and Mike Aunchman, Ed Bartholomew, Les and Kitty Benjamin, Neil Blackmer, Barton and Mary Campfield, Joe Capron, Mr. & Mrs. Homer Combs, Susan Cook, Leo Dewey, Robert Diekel, the late Harry Diekel, Joanne Farley, police officer James Fiorini, police officer Dave Gebo, former police officer Dan Gordon, former police officer Brian Gosselin, Darrin Gosselin, former Whitehall Police Chief Wilfred Gosselin, Susan Hallenbeck, police officer Timothy R. Hardy, Charles Kader, Ann and John Keys, Barbara Knights, Barry Knights, Whitehall Police Chief Richard LaChapelle, former police dispatcher the late Robert Martell, Dave and Marjorie Mohn, police officer Jim Murphy, Pat Norton, Fred and Carol Palmer, Travis Potter, Francis and Cheryl Putorti, Don Rogers, Police officer Thomas F. Ruby, Tom Rushia and Naomi Van Guilder, Dee Russell, Vernon Scribner, Frank Soriano, Cliff and Pat Sparks, Neil Spears, the Whitehall Village and Town Boards.

We wish to thank the following members of the media: David Blow, Brian Dow, the late Barney Fowler, the late Reverend George Greenough, Walter Grishkot, Doug Hajicek and the crew of Whitewolf Entertainment, Chris Hamilton, Barbara Hoag, Anna Lee, Gordie Little, Rob McConnell, Ian McGaughey, the late Don Metivier, Bill Miller, Mark Mulholland, Paul Rayno, Jeff Rense, Patrick Ripley, Joan Ritchie, Rusty Trombley, Darrin Youker, Jason Rowe.

We are grateful for the contributions of the following artists: Rob Dubois, Sharon Ellis, and Domingos Joaquim. A very special thanks to Eric Miner at Dungeon Art Studio.

Thank you to anyone not mentioned who helped in this project, and those who shall remain anonymous.

Foreword

I am delighted to be asked to write the foreword for this book. I regret that that honor could not go to my late friend, Richard Greenwell (founder and secretary of the International Society of Cryptozoology), who spent many weeks in the field in northern California (including the weeks shortly before his death) trying to obtain hard evidence for Bigfoot. I am also embarrassed to admit that I had no idea that Bigfoot sightings existed in New England. I must be more observant next time I'm riding my horse in the Connecticut woods! I think that most people believe that Bigfoot encounters are limited to the Pacific Northwest, so this book is an excellent addition to the literature on this enigmatic beast.

The majority of the book consists of descriptions of reported sightings of Bigfoot, divided by area (New York versus New England) and historical period (early evidence from the nineteenth and early twentieth centuries, including Native American lore and sightings by early settlers, modern day reports from the late twentieth century, and then some recent reports from the past few years). This wealth of reports provides fascinating reading, and while at times the reading of these reports can feel a bit repetitive, this very repetition is the core of the book's value. On the East Coast alone, where most people are unaware of Bigfoot sightings, there are dozens of documented cases (hundreds if you include the actual number of witnesses) from the past 500 years. These cases are usefully summarized in chronological order at the end of the book.

Skeptics often dismiss eyewitness accounts as "anecdotal evidence," as if this means this type of evidence has no use whatsoever in a scientific evaluation of the data. A recent example is Benjamin Radford in *Skeptical Inquirer*, May/June 2007, in his review of the Bigfoot book by Jeff Meldrum. Radford claims that there is simply no "good, hard evidence for Bigfoot," but why should "good" and "hard" be irrevocably linked in this fashion, implying that eyewitness accounts are no type of evidence at all? It may be true that many supposed Bigfoot sightings actually represent bears, but it's hardly scientific to draw from this the conclusion that all of the sightings represent bears! Let's be really scientific here: what is the statistical likelihood that all Bigfoot sightings are actually those of bears?

The value of the evidence in this book by the Bartholomews is the repetition of sightings under different circumstances, not only at a distance and/or in the dark, when a bear might be the culprit, but also close up and in broad daylight. I was particularly amused by the report in 2004 of an orangutan-like animal by visitors from Hong Kong — people who were quite unaware of Bigfoot legends so could hardly have had the preconceived mindset to imagine such a creature. Of additional importance is the constancy of reported details that are quite unbear-like, such as a pointy head, reddish fur, long strides while swinging the arms, etc., a height of seven to eight feet (much taller than a black bear), and the presence of human-like buttocks.

As the authors discuss, many of the large mammals that were discovered in the late nineteenth century, such as the gorilla and the okapi, were known first from eyewitness accounts, which were initially disbelieved. So why are skeptics today so quick to dismiss sightings of Bigfoot as providing any evidence whatsoever? I also think that most people are unaware of how elusive big mammals can be in the wild. Even in the open plains of the Serengeti, very few tourists get to see the rarer mammals such as cheetah and wild dogs. Yes, sightings are subjective, subject to fabrication and alternative explanations, and not repeatable under controlled circumstances; but, as science is a human endeavor, such accusations apply to many aspects of scientific evidence, even hard facts. Most types of evidence are subject to human interpretation, although I admit that a Bigfoot corpse would be fairly unambiguous.

A brief aside: When I first met my late husband, Jack Sepkoski, we engaged in a playful argument about the nature of paleontology as a "hard science." "A fossil is the ultimate repeatable experiment," I claimed. "Open the drawer, there's the fossil, close the drawer, open it again, there's the same fossil again." "Aah," said Jack, "but *is* it the same fossil?" At that point I knew that I'd met my intellectual soul mate! Somebody who, like me, dealt in scientific facts, but who also realized that even supposed "hard evidence" can be subject to different interpretations, and that the "literal truth" is, like Bigfoot, an elusive beast.

So, yes, it would be nice to have "hard evidence" of Bigfoot, but the absence of such evidence does not mean that no evidence exists. Eyewitness evidence may not constitute proof, but it is still important information. I am happy to dwell in the shadowlands of thinking that there is a lot of good evidence for the existence of Bigfoot, in terms of the plethora of eyewitness accounts, but I don't feel that I have to stake my claim on either side of the "believe in it or not" fence, which is a stance that many scientists seem to feel that they must take as card-carrying members of their profession. I'm not sure why so many scientists paint themselves into this type of corner. A testable hypothesis of any type contains many "maybes."

In the last chapter of the book the authors consider Bigfoot as a scientific phenomenon. I'm glad that they dismiss paranormal explanations, such as Bigfoot coming from a parallel universe during particular phases of the earth's magnetic field, or being some type of psychic projection. However, they do a careful consideration of these hypotheses, and it *is* strange how often reports say that the creature, after being sighted, just seemed to vanish into thin air. They also briefly consider evidence such as footprints and hair, as well as the possibility of hoaxes and people's overactive imagination.

Some of the "scientific" arguments presented against Bigfoot seem strange to me, at best. Colin Groves (a member of the International Society of Cryptozoology as well as a bona fide scientist) claims that there is no native North American primate for Bigfoot to have descended from, an argument that can easily be dismissed by noting the large numbers of mammals to immigrate to North America from Asia during the Pleistocene (including, of course, *Homo sapiens*). I rather doubt, as suggested by Warren Cook, that Bigfoot represents a surviving Australopithecine. Later Australopithecines, at least, were savanna animals, and had they ventured outside of Africa it is likely that we would have found their fossil remains there. Additionally, we share our mode of bipedality with Australopithecines, whereas reports of the locomotor style of Bigfoot (including analysis of the Patterson film) suggest a rather different mode of bipedal walking to the human condition. Note that it is not unreasonable to propose an independent origin of habitual bipedality in another primate. Apes in general have a tendency to locomote bipedally, in the trees as well as on the ground (see the recent paper in *Science* by Robin Crompton and his associates at the University of Liverpool). In addition, a late Miocene ape from Europe, *Oreopithecus*, has been interpreted as being fully bipedal, but in a completely different fashion from *Homo sapiens*, with short legs and a broad, tripodal foot (obviously not adapted for long-distance walking or running).

Another strange observation (from V. Rae Wigan) is that Bigfoot would be unable to survive in the boreal forests in the winter, especially as there is no evidence of tools. So how, then, do bears survive in the same habitat? In fact, Bigfoot's huge size is actually commensurate with it being a boreal herbivore; a larger-sized mammal has relatively lower metabolic requirements (i.e., less cost per gram of tissue) than a smaller one, and so larger mammals are able to subsist on low quality food. A horse can subsist on coarse hay, but a small deer would be unable to eat enough of this diet per day to survive. So, a prediction that one would make of an ape that had adapted to a boreal forest environment is that it would be considerably larger than tropical apes. This, of course, is not evidence that Bigfoot exists, but does show that various aspects of its supposed anatomy (also remember the large buttocks!) do actually fit with what one would expect of such an animal if it did exist. (I am at a loss to explain the pointy head, however!) This argument is not, of course, any type of proof for Bigfoot's existence, but I find it reassuring that its known features make it a biologically credible animal. This would be in contrast, for example, to something like a dragon, where the laws of physics dictate that the wings would be too small to generate flight.

There does remain the problem of the lack of hard physical evidence for Bigfoot, although this may not be quite as puzzling as one might imagine. Richard Greenwell told me that there are no bear bones found in the wild either, and that bear skeletons in American museums are all from captive specimens. Unfortunately I have been unable to verify this statement, although I consider it to be a distinct possibility. It may be true that there is no fossil record evidence of a Bigfoot-like creature, but it is also true that there is virtually no fossil record evidence of any living great ape (orang, chimpanzee and gorilla) in general, most likely because they inhabit forest environments with little opportunity for preservation; the more copious remains of hominine fossils (*Homo sapiens* and close relatives such as *Australopithecus*) are from dryer habitats. Note that the remnants of *Gigantopithecus*, a proposed ancestor for Bigfoot, are few and fragmentary, consisting largely of teeth. Thus arguments that if

Homo sapiens left dozens of fossils in North America then Bigfoot would have done as well are specious, as humans of the past 10,000 years in North America clearly lived in variety of habitats, including ones where fossilization would be likely. Also, early humans were clearly very numerous, increasing their chance of fossilization, while it's likely that a more specialized, boreal animal (as proposed for Bigfoot) would have had smaller populations. Note, also, that while there are some Pleistocene and Holocene fossil sites from New England and the Pacific Northwest, these are relatively few in comparison with the Great Plains area of the US. Lack of physical evidence for Bigfoot is problematic, but there are possible explanations for this that are not overly circuitous.

However, I wish that I could share the confidence of the authors that, if Bigfoot is real, it will inevitably be discovered by scientists. It has escaped detection so far, and there is hardly a plethora of people out in the field actually looking for evidence — I'm fairly confident that the National Science Foundation would not fund such a proposal. Additionally, with the encroachment of humans on the habitat of other large mammals, I consider it highly likely that, if Bigfoot did truly exist, population sizes would now be severely reduced and it would likely be in danger of extinction. It would be interesting to see if the number of sightings, relative to the number people in the area, have actually decreased over the past few decades (although this would presuppose evenness of reporting through time, which cannot be evaluated). However, we can still have hope. The existence of Bigfoot is, after all, a testable hypothesis, and thus remains within the realm of scientific enquiry.

A final comment about the "scientific" (or not) nature of the quest for Bigfoot. In the aforementioned issue of the *Skeptical Inquirer*, another writer, Michael Dennett, accuses Jeff Meldrum of performing pseudoscience because he supposedly cherry-picks data, misinterprets evidence, and fails to mention alternative opinions by other people. In actual fact, what Dennett is accusing Meldrum of doing is *bad science*, not pseudoscience, as such criticisms could be leveled at many people doing hard science such as physics and chemistry. Pseudoscience is when the principles of science are applied mistakenly or with deliberate intention to deceive, when claims of rigor are made but various scientific essentials are missing, such as the necessity for a testable hypothesis. "Creation Science" falls under this rubric. There is no way to falsify the hypothesis that something was designed by God (or by some other intelligent being). However, accusations of a similar mode of thought could be leveled at those skeptics who have already decided that it is impossible for Bigfoot to exist, so they either refuse to examine the evidence, or claim that it is all inadmissible, claiming arguments such as one proven hoax means that all sightings are hoaxes. Let's hope that a truly scientific approach to the existence of Bigfoot will one day yield different types of evidence from those that already exist.

CHRISTINE MARIE JANIS
Professor of Ecology & Evolutionary Biology
Brown University
Providence, Rhode Island

Professor Janis holds a Ph.D. in biology (Vertebrate Pateontology) from Harvard University (1979) and has an enduring interest in Cryptozoology, the study of animals that have yet to be proven, or are thought to be extinct.

Introduction

> ...the mysterious. It is the fundamental emotion which stands at the cradle of true art and true science.
>
> —Albert Einstein[1]

Could a hulking seven-foot-tall, human-like creature roam the green, fertile valleys and lush forests of New York and New England? It sounds like science fiction. After all, where are the fossils and bones? What about DNA evidence? With an army of hunters, hikers, campers and leaf-peepers scouring the woods each year, it's difficult to believe that someone has yet to shoot one or find the body of a dead or dying creature. Throughout human history, not one person has produced proof of its existence. From this standpoint, common sense tells us that Bigfoot is a myth.

Yet, here is the conundrum: how do we explain the hundreds of credible eyewitness accounts across New York and New England dating back to Indian lore and early European settlement? Most are respectable citizens who shun publicity. Numerous reports have come from police and conservation officers, many at close range. What about hair that belongs to no known primate, and trails of huge footprints found in the middle of nowhere? Why would someone go to such lengths, not knowing if anyone would visit the area for months or years, and their efforts could be obliterated in a matter of minutes or hours by rain or wind? This is the great paradox which cuts to the heart of the global mystery, and applies no less to New York and New England. Science and logic would suggest that Bigfoot cannot exist, yet it seemingly does — and not just a few sporadic sightings in a handful of remote areas. This state of affairs has led some researchers to postulate an array of esoteric theories, most notably, that Bigfoot is a paranormal entity. However, naturalistic explanations seem more plausible. A chameleon changes color to blend with its surroundings and become seemingly invisible. The peregrine falcon has remarkable eyesight enabling it to spot small animals up to five miles away. Owls have developed incredible night vision. The sooty tern can stay aloft for more than three years without landing and sleeps while it flies. Is it too far-fetched to think that Bigfoot could have evolved unique defense mechanisms that allow it to elude capture?

Bigfoot is an undeniable part of the history and folklore of New York and New England. It is also part of an enduring global enigma. For when most people hear the word Bigfoot, they think of the Pacific Northwest and the long history of Sasquatch sightings in Northern California, Washington, Oregon and British Columbia. For others, the word conjures up images of a yeti or abominable snowman trekking on a lonely, desolate mountain slope in the Himalayas. This book will change these images to include the northeastern United States.

Yes, it sounds like science fiction, but there is a reason why this book appears on nonfiction shelves: an impressive body of evidence from credible witnesses who believe they saw 'Bigfoot.' Readers are invited to weigh this evidence and make up their own minds.

ROBERT AND PAUL BARTHOLOMEW

Whitehall measures both in U.S. Navy history and Bigfoot lore, as will be revealed in this work. The Bigfoot carving seen under construction and completed (right), is by Domingos Joaquim. It was donated to the town of Whitehall and now stands outside the town's pavilion.

PART 1 BIGFOOT Encounters in New York

1 The Early Evidence: First Reports – 1949

One measures a circle beginning anywhere.
—CHARLES FORT

Indian Lore and Pioneer Days

The words "New York" conjure up images of majestic skyscrapers, bustling airports, monumental traffic jams, crowded streets, and a smog-shrouded skyline. Yet, many people will be surprised to learn that much of the State harbors small towns and sleepy villages nestled in sparsely populated mountains, forests and valleys, several of which have been Bigfoot hotspots for as long as residents can remember.

The most impressive body of evidence that a man-like creature inhabits the region, can be found in the picturesque Adirondack Mountains in the State's northeast, which cover 12,000 square miles. In the heart of these ancient mountains lies the Adirondack Park encompassing six million acres of protected wilderness.[2] It isn't far-fetched to think that such a creature could make its home in the region. Sightings date back to Indian times. The existence of a man-beast residing on these lands was debated among the earliest European settlers.

The word "Adirondack" means bark eater. It is of Iroquois origin and, surprisingly, may have been used as a derogatory term to describe a neighboring band of rival Indians — the Algonquin. The

Algonquin told stories of the *windigo* or "giant cannibalistic man" who, according to legend, roamed the countryside. Its diet would certainly not be on the menu of any fancy New York restaurants; it was said to consist of meat, wild mushrooms, swamp moss and rotten wood. The creature was reputed to have a black mouth, no lips and made hissing sounds and reverberating howls "like that which grouse make when they drum."[3] Similar noises are attributed to Bigfoot today. The windigo legend could be found throughout all Algonquian-speaking people. They believed that the windigo (also spelled wendigo, weedigo, witiko) had the capacity to "live unprotected in a harsh physical environment, easily find food, and withstand isolation…"[4] One modern-day Native American account of the windigo, describes it as "a giant thing, swift…and covered with hair, and has eyes like two pools of blood. And there's this smell, like rotting meat."[5] Once again, this description is similar to Bigfoot reports today.

The Iroquois have a similar oral history of flesh-eating Stone Giants who used rocks as weapons and possessed powerful physiques that were capable of uprooting small trees.[6] In Western New York, Iroquois tribes referred to it as *Ot-ne-yar-hed* or Stonish Giants.[7] Members of the Iroquois Confederacy, such as the Seneca of Western New York, have a similar legend of mighty, cannibalistic Stone Coats, while the Onondaga of north central New York referred to these creatures as the Northern Giants.[8] Just across the Canadian border along the upper St. Maurice River in Quebec Province, the Algonquian-speaking *Tete-de-Boule* (Attikamekw) called such creatures *Kokotshe*. Writing in the journal *Primitive Man*, anthropologist John Cooper said the Tete-de-Boule did not consider these beings to be literally made of stone. "He used to rub himself, like the animals, against the fir, spruce, and other resinous trees. When he was thus covered with gum or resin, he would go and roll in the sand, so that one would have thought that...he was made of stone."[9] Perhaps this was a way to keep insects at bay, provide camouflage or mask one's scent.

The first European to write about such creatures in what is now northern New York, was the French explorer Samuel de Champlain (1567–1635) for whom Lake Champlain is named. In 1603, while voyaging on the St. Lawrence River, he wrote that so many of the local Native American tribes recounted stories of a huge, hairy man-beast who was said to walk the woods, that he believed there must be some truth to the tales of the *Gougou*.

A translation of Champlain's French logs from his voyage of exploration on the St. Lawrence in 1604, reveals a lengthy reference to these mysterious creatures. "There is another strange thing worthy of narration, which many savages have assured me was true; this is, that near Chaleur bay, towards the south, lies an island where makes his abode a dreadful monster, which the savages call Gougou..." The natives believed that many missing persons had been eaten by these creatures, though it appears that no one saw it in the act: Champlain said that many a native swore to him they had seen a Gougou. He said that fellow countryman and explorer Sieur Prevert of St. Malo told him that during a voyage in the region, "he passed so near the haunt of this frightful beast, that he and all those on board his vessel heard strange hissings from the noise it made, and that the savages he had with him told him it was the same creature, and were so afraid that they hid themselves wherever they could, for fear it should come to carry them off." Champlain was convinced that the natives were not just telling tall tales, but that the region was the home to the strange creature. "And what makes me believe what they say, is the fact that all the savages in general fear it, and tell such strange stories of it...[and] I hold that this is the dwelling-place of some devil that torments them in the manner described."[10] Some accounts of the Gougou that were told to Champlain by the natives, were clearly exaggerated, such as claims that the creatures were taller than his ship's masts.

The 19th Century: Man-Beasts and Wild Men

In 1818, a mysterious Bigfoot-like creature was spotted along the western fringe of the Adirondacks in the Jefferson County town of Ellisburgh, a suburb of Watertown on the remote Canadian border. The incident took place in late August and involved "a gentleman of unquestionable veracity" who said he saw a wild man dashing through the woods. The creature reportedly came to within twenty yards of the man. After staring at him momentarily, the man-beast ran off. The witness said its body was distinctly bending forward as it fled. The creature was "hairy, and the heel of his foot narrow, spreading at the toes." The sighting triggered a massive hunt involving hundreds of locals who split into several search parties.[11] Despite these extraordinary efforts, they failed to uncover so much as a trace. It was as if the creature had simply disappeared from the face of the earth.

Twenty years later during the summer of 1838, a boy reported seeing what may have been a juvenile Bigfoot covered in black hair, in New York's southern border region. The encounter came to light after his father recounted the incident to a journalist for the *Dorchester Aurora* newspaper. The encounter occurred on or about August 13, in the town of Silver Lake in northern Susquehanna County in Pennsylvania. Silver Lake borders New York. The newspaper article stated:

> The boy was sent to work in the backwoods near the New York State line. He took with him a gun, and was told by his father to shoot anything he might see, except persons or cattle.
>
> After working for a while, he heard some person, a little brother he supposed, coming toward him whistling quite merrily. It came within a few rods of him and stopped.
>
> He said it looked like a human being, covered with black hair, about the size of his brother, who was six or seven years old. His gun was some little distance off, and he was very much frightened. He, however, got his gun and shot at the animal, but trembled so that he could not hold it still.
>
> The strange animal, just as his gun "went off," stopped behind a tree, and then ran off, whistling as before. The father said the boy came home very much frightened, and that a number of times during the afternoon, when thinking about the animal he had seen he would, to use the man's own words, 'burst out crying.'
>
> Making due allowance for frights and consequent exaggeration, an animal...has doubtless been seen. What it is, or whence it came, is of course yet a mystery.[12]

In 1868, there were several sightings of a wild man near the tiny rural community of West Milton in Saratoga County, about ten miles southwest of Saratoga Springs. Attempts to capture the creature were unsuccessful.[13] The following year, a strange animal described as half man, half monster, caused a sensation in the vicinity of Woodhill and Troupsville in southern Steuben County in southwestern

New York. Between mid-June and mid-July 1869, at least 100 residents reported seeing the creature along roadsides, in open fields and in the woods. Despite a thorough search, it was never found. During the scare, many parents stopped sending their children to school after the "strange being rapped on the windows of the school house," frightening the students inside. On the 12th, a hunting party of 200 men was organized to track down the wild man. Upon spotting it on the outskirts of a wooded area, the posse went in hot pursuit, getting to within forty yards, at which point it gave out an "unearthly shriek." One of the leaders said he raised his gun and took aim, but couldn't bring himself to fire the weapon on account that the creature looked so human. "I drew my rifle, intending to halt him or send a bullet crashing through his skull...[H]e sprang with the agility of a deer toward the woods. I did not fire, because on second thought I doubted my right to take the life of any human being, however wild, until he had at least violated some law." He said the creature moved about "with a springing, jerking hitch in his gait, [which] gave him more the appearance of a wild animal than a human being." The witness said the sight of the creature remained firmly etched in his mind. "The long, matted hair; the thick, black, uncombed beard; the wild, glaring, bloodshot eyeballs, which seemed bursting from their sockets; the swage, haggard, unearthly countenance; the wild, beastly appearance of this thing, whether man or animal, has haunted me continually by day and night..."[14]

The following month another wild-man scare took place in extreme northern New York near Sucker Brook, a tiny hamlet one mile south of Ogdensburg on the Canadian border. Some descriptions suggest that it was a hermit, but others said it looked more like an animal. For instance, some witnesses said its arms and legs "were covered with long hair." One boy said the wild man took about a dozen minnows from his pail and ate them.[15]

In March 1883, there is a curious report from Port Henry, just north of Ticonderoga on Lake Champlain. According to the *Plattsburgh Sentinel*, "The great scare at Port Henry is the wild man, who scares women and frightens the children." Some witnesses said it appeared to be wearing "an overcoat." Could this have been fur or hair that was mistaken by witnesses who were trying to make sense of what they were seeing and assumed it must have been a human? Despite numerous sightings, "it" was never found.[16]

Between August and November 1883, there were numerous wild-man sightings in extreme south central New York near the small town of Maine on the western border of Broome County, northwest of Binghamton. As a result, the gathering of wild berries, a livelihood for many local families, "was entirely neglected in some parts of the neighborhood." The elusive creature was first spotted by a group of berry pickers, running from a clump of bushes and into the woods. It was described as "low in stature, covered with hair, and running while bent close to the ground." During September, two workers at Maine's Sherwood Tannery were loading bark in Lewis's woods near Newark Valley Road, when they heard a frightening shriek from the nearby bushes. The startled men said the creature let out another piercing cry before it "ran off like lightning." They also noted something distinctly odd: the hairy man appeared to have no forearms as its "arms ended at the elbows." Nightly shrieks attributed to the wild man continued to be heard in the following days, prompting twenty men to organize a hunting party to track it down, though nothing was found.[17]

In early November 1888, there was considerable excitement among residents living in the vicinity of Big Run, in the rugged north central part of the state in Jefferson County. Huge footprints were discovered that were attributed to a wild man. The human-like prints measured eighteen inches in length.[18]

There were sightings of a wild man near Rockaway on the western tip of Long Island during late November 1895. One witness, John C. Ennis of Far Rockaway, said that as he approached the figure who was standing on the sand, it "gave a piercing shriek and plunged into the surf, disappearing from sight." The cry was so powerful that it spooked a team of horses who couldn't be stopped until reaching nearby Eldrat's Grove.[19]

An Incredible Tale

Two years later, an incredible account emerged from the tiny Delaware County resort town of Margaretville in the eastern part of the state near Newburg. On the night of July 26, 1895, Peter Thomas was driving a team of horses on a remote stretch of road when "a wild-eyed man or ape" sprang from the bushes and stood in the road ahead. Mr. Thomas said he was dumbfounded at the sight and temporarily froze as he looked at the creature which was "brandishing a pair of long and hairy arms and uttering a raucous, inarticulate cry." He said the ape-man grabbed and twisted the neck of one of his horses, then dragged it off into the darkness. The creature was described as "seven feet high, of human shape, covered with hair."

The next day local farmer John Cook reported that he shot at and apparently struck "a ferocious ape-like being," after which he said, the creature came at him and threw him to the ground. Cook said the wild man stood "about seven feet tall, entirely nude, covered with black hair, with a long beard and with teeth which project from its mouth like fangs." Later that day a group of farmers and young boys went searching for the man-beast. The remains of Mr. Thomas' horse were reported to have later been found nearby, along with the splintered bones of sheep and cows.[20] The incident sparked a debate as to whether it was humane or even legal to hunt down such a creature. A reporter for the *Illustrated Buffalo Express*, observed: "A party has been organized to hunt him out and slay him. But the question has arisen whether they have the right to kill a wild man. Some say that it would be murder" under the belief "that wild men are protected by the game laws."[21]

A Surge in Downstate Reports: 1890–1930

There is evidence to suggest that urban sprawl and dwindling habitats forced these creatures to slowly migrate to the north and west. Between 1890 and 1930, there were a series of sightings downstate on the outskirts of populated areas. In late autumn 1893, "Red" McDowell and George Farrell were paddling along the remote Rockaway Inlet, six miles from Rockaway in eastern Long Island, when they spotted a wild man standing on shore. As they rowed toward it for a better look, the figure uttered "wild cries" which terrified the pair who quickly paddled off. The following evening of November 23, John Louth saw a creature of similar appearance while driving in Rockaway Park. The next day his eighteen-year-old daughter came face-to-face with a frightening animal which jumped out of the bushes, knocked her to the ground and ran off "uttering strange yells." It was described as "a wild man, large in appearance, with fierce, bloodshot eyes, [and] long, flowing, matted hair." The hair around its face was said to be shaggy. Some theorized that the creature was a sailor who had been missing and was presumed drowned after the schooner *Maggie Devine* ran aground in the area several weeks earlier. While it is possible that missing first mate James Rush had become delirious from exhaustion and lack of food, it seems far-fetched given that he was never found. One witness even saw the wild man munching on a raw chicken. It's hard to imagine that even a half-crazed sailor couldn't find a more appealing meal.[22]

Four years later in early September 1897, another police search was mounted in western Long Island, this time for a wild man with matted hair who was spotted by a dozen people in the vicinity of Bronxville. From descriptions, it is almost certain that the figure was an escaped convict, though there is no account of him being killed or captured.[23] A wild man resembling a gorilla, was spotted in April and May 1899, in the small village of Johnsonburg in Wyoming County of Western New York. Sightings of the creature continued for several weeks in spite of efforts to hunt it down. Missing chickens and sheep were linked to the creature's appearance. In one encounter, several women were strolling through the woods when they saw the thing. Upon spotting them, they said it "ran swiftly up a huge tree and was lost sight of." While many residents thought the creature was an escaped gorilla, this is difficult to believe as there were no reports of escaped gorillas, hunting parties failed to shoot it, and no body was ever found.[24]

At about the same time, hunters were scouring the woods for a wild man in the town of Dresden, thirty miles northwest of Ithaca in the Finger Lakes region of Central New York. Witnesses described it as a "strange person or thing." In one incident, a group of local farmers joined up with workers for the Fall Brook Railroad, converging on the scene of a recent sighting. When one of the pursuers walked through a clump of bushes, he reportedly fell over the wild man, who was said to be lying down. His description of the "thing" is nothing short of bizarre. The startled man said he thought the creature had nothing on but a red shirt — clearly indicating that it was a human being. As for the rest of its body, it resembled "a gorilla, being covered with a dark sort of hair or skin." The party of men, armed with guns, clubs and stones, chased the figure through the woods, but when they came out on the other side of the trees, it was gone—seemingly vanishing.

So, what did the man stumble across? A bearded hermit? A bear? Bigfoot? If it was a bear or Bigfoot, how does one explain the red shirt? Skeptics are fond of pointing out that human perception is notoriously poor and people often see what they expect to see. Could it be that the witness had expected to see a man, so his mind initially constructed a shirt? A reporter for the *Rochester Democrat and Chronicle* dismissed the gorilla and animal-like descriptions. "The unsightly appearance of the man is thought to have been greatly exaggerated, those who were engaged in the chase evidently being considerably excited, which, together with the darkness, probably made him appear worse than he really was."[25] He then went on to conclude that the wild man was an escaped convict, mental patient or tramp. This is a common pattern in early wild-man accounts and is based on popular expectations. Another common element is that despite massive, thorough searches of the area, he, she or it usually manages to get away. How should we interpret these sightings? Through the early twentieth century, there were a number of confirmed wild men — that is, people who had abandoned civilization for one reason or another, to live in the wilderness. These hermits were usually docile and wanted nothing more than to be left alone. Based on witnesses descriptions, others were likely escaped convicts or mental patients. Another challenge in interpreting these early reports has to do with the term "wild man," which was used to describe not only hermits, but Bigfoot-like creatures.

During the early twentieth century, sightings of large, ape-like creatures in the Long Island countryside and the suburbs of New York City became more common. Were they escaped pets? Could they have been relatives of modern-day Bigfoot who later migrated to more rural parts of Upstate New York and New England? If they were escaped pets or zoo animals, it is curious that no trace of these creatures was ever found — and none were reported missing. A wild-man scare swept through the modern-day home of the New York Yankees, the Bronx borough of New York City during July 1906. The creature was described as "a lean, ferocious-looking human covered with more hair than the law permits" and eyes that glowed in the dark. Sightings were concentrated in the town of Wakefield, where many residents were then fearful of venturing out after dark.[26]

In February 1909, a strange creature "with eyes of flame" and emitting a "blood-curdling shriek" terrified Long Island residents in Eastport, Westhampton, Patchogue and Quogue. While not one of the many witnesses got a clear view, several said it resembled a monkey or baboon. One theory held that it was a pet from a deep bark vessel that came to grief on the shore the previous fall. On February 7, a group of men with guns scoured the woods for the nocturnal beast but this and other attempts to capture or kill the mystery creature all proved unsuccessful.[27]

One of the few sightings from the Adirondacks during this period, occurred at Rainbow Lake, thirty miles south of Malone near the Canadian border. The wild-man scare took place in November 1909. Many chickens were said to have mysteriously vanished at around this time, and many people reported seeing the "long-haired man." As the Adirondack branch of the New York Central Railroad runs through the remote region, one theory held that an escaped convict or lunatic had jumped off a freight train and was hiding out in the wilderness. A number of residents who had sighted the wild man wrote letters to Franklin County Sheriff Steenberge, asking him to search for and capture him.[28]

The tiny village of Ellicottville in Cattaraugus County about forty miles south of Buffalo, was the scene of a wild-man scare during September 1915. Despite police searches, there is no record of the naked man or creature being caught. As is so often the case with these reports, the reporter assumes that witnesses were seeing a lunatic.[29]

During the autumn of 1921, there were two mystery creature flaps in the northern Adirondacks. During September, residents in the tiny towns of Bangor and Brandon in Franklin County, were on edge after several sightings of a wild man. The *Plattsburgh Sentinel* reported that despite the best efforts of hunters and trackers, "no one has succeeded in getting near him." Meanwhile, many parents were so concerned for the welfare of their children that they refused to allow them to walk home from school, instead, picking them up at the school door. Bangor is situated in the village of Malone, while Brandon is in a wilderness region southwest of Malone.[30]

In October and November 1921, residents near Malone were again living in a state of fear following another flurry of wild-man sightings. Authorities organized search parties under the direction of the county Sheriff. Was it the same creature? On several occasions it appeared to be holding a club. Most sightings were clustered around the hamlet of Skerry, twelve miles southwest of Malone. A reporter for the *Dunkirk Evening Observer* described the tense state of affairs: "Women sleep ill o' nights, children are kept from school, or guarded by adults on their way there and back, lonely females cower behind locked doors and men wag their heads in gossip as they ponder over the puzzle of the wild man..." While the residents may have been seeing a bearded, disheveled hermit, it is curious that he was never captured, despite numerous sightings.[31] Some Franklin County authorities considered the story unlikely, instead opting to believe that it was "a clever ruse effected by bootleggers to take advantage of the absence of officers," so they could temporarily smuggle liquor across the Canadian border with ease.[32]

A scare involving a "ferocious baboon" took place near Babylon, Long Island in the mid-autumn of 1922. By November 5, in an effort to relieve the tension and perception of danger, organized hunting parties weaved their way through the surrounding woods. While some said it resembled a baboon, others said it was more like a gorilla.[33]

Nine years later during the summer of 1931, sightings of a "wandering gorilla or perhaps a chimpanzee" were recorded near Huntington Long Island. The four-foot-tall creature was first spotted by six people near Mineola in June. Heavily armed police rushed to the scene thinking it may have been an escaped zoo animal, but no trace of the creature was found. Nassau County Police checked with area zoos but all animals were accounted for. By June 29 the scare reached a fever pitch as police captain Earl Comstock directed several teams of armed police accompanied by two dozen volunteers who scoured the countryside. The only possible sign of the creature: strange human-like tracks that seemed to be of an animal walking on two legs. Their size was about that of an adult hand, though the big toe was set further back than on a human. On July 18, a gorilla-like creature was spotted by a family in Huntington as it crashed through some shrubbery. Three miles away, a farmer saw a similar strange animal. Once again, police were stumped, searching in vain.[34]

The Saga of the Adirondack Wild Man

In the winter of 1932, two trappers encountered a Bigfoot-like creature roaming the backwoods of Hamilton County. The saga began one cold night in February, when two Indian Lake cousins, Richard Farrell and Reg Spring, were trapping for furs. They came upon a cabin two miles south of Blue Mountain Lake in the area known as the O'Neil Flow. As they grew closer they spotted a mysterious, hulking creature covered in hair, seven feet tall, peering through a window. When it ran off, they examined the footprints. They were massive, measuring thirty inches in length.

The frightened cousins contacted authorities and a posse, led by the New York State Police, was organized to search for the giant. Trooper Charles B. McCann headed the party which included fellow Trooper Addison Hall, conservation officer Jack Farrell, Charles Turner and Ernest Blanchard of Indian Lake, and the two trappers. The group headed straight for the cabin and soon spotted the huge footprints in the snow.

On the second day they found the creature in a cabin near Dunbrook Mountain in the town of Newcomb. After they surrounded the building and called out to him, a shotgun blast blazed from the cabin. The party returned fire, hitting the wild man in the leg, but he still managed to escape through a window. They were all surprised at how quickly he was able to dash over the snow. They now realized that the "creature" was a man. When they spotted him hiding behind a pile of logs, he yelled out, "I just want to be left alone. Go way." The man fired another blast, striking Turner in the hip and knocking him out of action, though it was only a flesh wound. A hail of bullets quickly sent the man crashing to the snow. As the men inched closer they realized that the wild man was a black man covered in several layers of untanned bear and deer skins. In fact, by the time they had removed the many layers, they found that the "giant" was just five feet, six inches tall and weighed a mere 160 pounds. The man's identity remains a mystery. All that was found in his possession was four dollars in Canadian currency. When Hamilton County lawmakers refused to hand over the $75 needed to bury the mystery man, his body was placed in an unmarked grave at Potter's Field in North Creek.[35]

The shooting caused an outcry and many residents believed the officers should have been charged with cold-blooded murder, suggesting that a gift of food and clothing would have been more appropriate. While an inquest into the shooting cleared those involved on any charges, it was also clear that he was neither wanted for any crimes nor escaped from a mental institution.[36] In retrospect, racism may have played a role in the shootout.

Amityville on Long Island is perhaps best known as the setting for the bestselling book, *The Amityville Horror*, in which it is claimed that a house owned by George Lutz and his family, was the scene of a series of hauntings.[37] Long before the book and subsequent film captured the imagination of the American public during the late 1970s, there were reports of a huge ape or demon haunting the community. On September 5, 1934, the *New York Herald Tribune* reported that Amityville authorities were searching for a mysterious ape-man who was on the loose, frightening residents under the cover of darkness. Some residents considered it to be a paranormal entity. The creature was first seen in late August. Then, in the early morning hours of September 4, it shredded several mattresses and a fur coat at the home of Mrs. Alfred Abernathy of South Amityville. It also scratched up one of their cars. As a result, the next evening it was reported: "Most of the male residents of the neighborhood are sitting on their porches waiting for the animal with shotguns, rifles, revolvers and garden hoses."[38] The mystery creature of Amityville was never caught. It was as if it had vanished into thin air.

During the 1940s and 50s, there was a lull in Bigfoot reports across the State, perhaps in part due to the focus on World War II and its immediate aftermath, the Cold War. These events dominated headlines and perhaps editors were in a more somber, less receptive mood compared to what many may have seemed frivolous compared to surviving Nazism and the subsequent nuclear standoff with the Soviets.

In mid February 1941, a wild man was reported near Copeland School in Potsdam. A letter to a local newspaper read in part, "Be on your guard for a wild man who chased Miss Marion Smith, after skating on Mr. Hanson's ice pond Tuesday evening near the Copeland School." The victim was said to have "cried with fear."[39]

In November 1948, a hunter came upon a wild man seven miles from Tupper Lake, near the deserted, heavily wooded village of Kildare. The encounter involved well-known fruit and vegetable wholesaler Lawrence Peets of Schuylerville in nearby Saratoga County. Peets said he was walking along a trail when he heard "whimpering." He turned around to see what looked like a man covered in rabbit skins. He said, "His hair was blonde and unkempt, and one arm was wrapped in grass and fern brakes. He carried a dead rabbit in the other hand, but had no weapons." Peets called out to the wild man but said it bolted off in fright. After joining up with fellow hunter Alfred Berard of North Tonawanda, the pair searched the area but could not locate the man or beast though they believed they could hear it whimpering from a nearby hilltop.[40]

2 Modern-Day Reports: 1950 – 2000

> I'm not saying this is a monster...but there is something out there, and it's no animal that belongs in the northern part of this state!
>
> —Wilfred Gosselin, Whitehall, New York Policeman [41]

In 1961, an incident took place at East Bay in Whitehall, nestled in the Adirondacks at the base of Lake Champlain in the northeastern part of the state. It would be the first in a series of several dozen reports to come from this small, rural farming community. Two young women were walking when they encountered a large figure that moved very close to them. "Oh my God. I'll never forget it," recalled one of the witnesses. The pair was terrified by their encounter. "I just screamed...and I just took off running," she said, until reaching her father's house, which took about two minutes. After recounting what had happened, her father telephoned police who searched the area but found nothing. While neighbors laughed off the incident as a prank by someone out to scare the two, the woman said that she and her companion were certain it was no one playing a joke. She said it was too tall and too big to have been a prankster.[42]

A Cone-Shaped Head in a Cabin Window

In the summer of 1969, four people were sleeping overnight in a small cabin at the Pumphouse campsite at Long Lake, in the heart of the Adirondacks in northern Hamilton County, when an ape-like creature was spotted peering at them through a window. A small oil lantern illuminated the inside of the cabin when the encounter took place at 11:10 p.m. One of the men, identified as J.G., said his wife told him she could see a raccoon staring at them through the window at the back of the cabin. Rolling over in bed, he glanced up and saw a large cone-shaped head and a dark face that appeared to be pushed in. Brownish fur encapsulated the face. The next morning at a nearby stream, the couple found what appeared to be a heel print eight inches wide.[43]

In 1970, a forty-year-old local school teacher named Susan was driving between Glens Falls and Corinth, New York, a distance of some ten miles, when a creature darted out in front of her car. She said the figure was "big and hairy" and stood six to seven feet tall. Susan was accompanied by a friend who confirmed the incident. The creature was whitish in color with "big thighs" and "huge legs." "It was furry" and ran upright. She said that "it had a bumpy head—pointy, and hunches when…running."[44]

One of the thickest wilderness areas bordering Whitehall is the Galick Nature Preserve located in rattlesnake country. The region is virtually untouched by the hand of modern man. The terrain is crossed by a series of narrow dirt roads which are treacherous during the best of times, and impassable each spring. In the 1970s, one of these creatures was apparently brazen enough, or simply curious enough, to walk onto the porch of a resident there. The man told friends that he had both seen and heard a large, ape-like creature.[45]

On the night of February 16, 1974, a man was enjoying some quiet time with his girlfriend after parking on the side of Winters Road in East Whitehall. At about 10:30 p.m., he was startled by a dark form silhouetted against the background snow. He estimated that the figure stood at six-and-a-half to seven-and-a-half feet tall, with a "large body mass" that was moving in "an effortless shuffle at about twenty miles per hour." The man said he spotted the creature sixty-five yards away, but never told the girlfriend. He said he "made up an excuse about having something to do in town and promptly left." He added that the experience made him edgy. "It was eerie to say the least," he said.[46]

Between December 1974 and January 1975, there was a sighting flap in the vicinity of Watertown on the northwestern fringe of the Adirondacks. On the afternoon of December 7, two boys, ages eleven and twelve, were walking up a knoll behind the St. Andrew's Episcopal Church parking lot when they were startled by a "loud roar." The pair turned around and saw a creature that stood six feet tall with its arms raised above its head. They fled. A police search of the area turned up nothing.[47]

The following month three people spotted a five-foot-tall creature that was "just walking and swinging its arms" as it made its way along State Street Hill. Then, on January 21, 1975, a big hairy animal that was estimated to stand five feet nine inches tall, was seen by a nurse driving near St. Andrews Episcopal Church at 10:45 p.m. Mrs. D. Daly said the creature didn't seem to pay any attention to her—as if she weren't there. It was walking from the church parking lot to the entrance of the churchyard. She said the creature had long hair and its arms hung down just ahead of its body. A set of ten-inch-long footprints with just four toes, were found in the vicinity.[48] In the early morning hours, a couple said they saw a bear-like figure in the parking lot.[49] An extensive swampy area and a trash dump are situated behind the church.

Activity soon shifted back to Whitehall on the eastern edge of the Adirondacks. Late one spring night in May 1975, Clifford Sparks, a former dairy farmer who converted his stead into a modest golf course, encountered a "sloth-like thing" while maintaining the greens on his rural Skene Valley Country Club. The encounter happened at 11:30 p.m. after Sparks said he had driven out to the first green in an electric golf cart. That's when he said "this great big hairy thing was silhouetted against the moonlit skyline." The creature was very tall, had a huge girth and the sight frightened him. "The hair on the back of my neck stood right up, because these carts don't go very fast," he said. Yet Sparks also made a curious observation: despite its size and obvious strength, it seemed more startled of him than he was of it.

Clifford Sparks

Sparks said that whatever the creature was, it wasn't graceful. "It lumbered and walked with a very clumsy gait…The creature had a different knee and leg action than a man…I don't know how, just different." The head was cone-shaped and "it didn't seem to have much of a neck." The shoulders were not broad. He said it was no more than thirty-five feet away, then ran off, crashing through a nearby wooded area in an awkward manner.[50]

The following month, seventy miles to the north, two men said they spotted "Bigfoot" squatting along the side of Route 3 in Saranac Lake in the heart of the Adirondacks. The figure then stood and walked off, disappearing into the nearby bushes.[51]

On August 11, 1976, two teenage boys told of two encounters with a huge creature in woods a few miles east of Watertown. They were walking on Overland Drive at 5:00 a.m. when they heard a series of screams and thumping noises that persisted for fifteen minutes. At 5:45 a.m. they spotted a black, hairy creature in the road ahead, standing on two legs. When they yelled in its direction, it ran off. In the morning they saw it again, describing it as having broad shoulders and being eight feet tall.[52]

Sketch of the creature seen by Clifford Sparks, by Rob Dubois.

As a result of many Bigfoot sightings in the Skene Valley area, the local country club now uses an image of the creature as its logo.

Face-to-Face with Police

The best documented and most spectacular series of sightings in the region occurred in rural Whitehall on the night of August 24, 1976. Whitehall teenagers Martin Paddock, Paul Gosselin and Bart Kinney told police they saw a seven- to eight-foot-tall, brown, hairy creature in a field off Abair Road. When television reporter Tracy Egan from Channel 6 in Schenectady, asked Paddock on camera what he did after he saw the creature, he replied, "Well, there's about sixty feet of black rubber in the road to show you what I did."[53] The next day farmer Harry Diekel, who lives within a mile of the first sightings, found "big, human footprints" in a nearby field. A ravaged deer carcass was also found nearby. Diekel said that a similar creature was spotted in the area in 1959.[54]

At 10:00 p.m., Paddock and Gosselin were in a pick-up truck when they spotted a hulking human figure with powerful, muscular shoulders, standing on the side of Abair Road. Paddock turned around and went back to get a closer look, parking on the roadside. They were soon spooked by a strange noise that sounded like a cross between a woman screaming and a pig squealing. Suddenly, they spotted a huge, hairy, ape-like creature in a field seventy feet away. It was at this point that Paddock sped off to get the police, who were understandably skeptical. That's when they picked up their friend Bart Kinney and the three returned to the site and saw it again.[55]

Gosselin said the creature was seven to eight feet tall as it stood on a small knoll near the shoulder of Abair Road. It was muscular yet stocky, covered in coarse brown hair and weighed about 350 pounds. He said the eyes were large and reddish but didn't glow in the dark. Kinney said that the creature walked slowly, its shoulders stooping, and was moving on two legs.[56]

The next night, Paul's brother Brian, an off-duty Whitehall Police officer, went to the site with a New York State Trooper to see if they could spot the creature. At around midnight, Gosselin was turning his police car around in the middle of Abair Road, when he spotted a pair of red eyes reflecting off his headlights. He quickly shut the lights off and radioed the Trooper what he saw. The Trooper shone his spotlight on a nearby hedgerow, at which point Gosselin said he heard the sound of something crashing through the brush and trees so he turned his headlights on. Just thirty feet away was a seven-and-a-half- to eight-foot-tall creature that he estimated

Drawing by Eric Miner from witness descriptions of the creature sighted along Abair Road in August 1976.

Drawing of the Abair Road creature by Sharon Ellis from witness descriptions.

weighed 400 pounds. Gosselin got out of his car, knelt down and aimed his .357 magnum service revolver directly at it and prepared to shoot, but his finger froze. He said he couldn't bring himself to fire on the creature as it looked too human.[57]

Gosselin said the creature stood there for a minute before vanishing into the night, affording him a good look. It had "big red eyes that bulged about half an inch off its face. As far as the mouth and nose, I didn't notice anything 'cause I was too scared — shook up...'cause I was staring at the eyeballs...As far as the ears go, it has no ears, and it has no tail...and it doesn't walk on all fours, it walks on two like a man would."[58] Gosselin said, "It's covered with hair, dark brown, almost black and the back end of it, the hair was more or less wore off because you can see the cheeks of the buttocks sticking out through the hair that was more or less worn. He was covered with clay on the backside. His arms hung just about eight to ten inches below his knees. He walks with a hunch and it didn't run, although it could move fast."[59] The creature then ran off screaming as the Trooper shone a light on it.

That autumn, Whitehall village police sergeant Wilfred Gosselin said he was hunting near the intersection of Abair Road and Route 22A when he heard a frightening "eerie high-pitched yell." He said the sound was unforgettable and persisted for nearly a minute. Gosselin said a herd of nearby cows stampeded out of the field while his hunting partner, his brother Russell, "came out of there as white as a sheet."[60] Gosselin was a well-known and respected officer on the police force and would later be named chief.

Just a few days later, another spectacular encounter was recorded in the area. Police are known for responding to unusual calls, but the evening of September 1, 1976 proved to be exceptionally strange. A man from the nearby town of Granville ran into the Whitehall Police Station and excitedly proclaimed, "I shot Bigfoot." The official police log for that night read as follows: "Frank McFarren of Granville came and reported that at 11:10 PM he shot four .12 gauge rifle slugs and six to eight .22 rifle bullets at a huge creature that came at him at C Falls Road. Notified N S police." Dispatcher Robert Martell said that McFarren seemed sincere and very scared. He was escorted back to the site by New York State Police and off-duty Whitehall police. When they reached Carvers Falls Road, in remote north Whitehall, they found a spent shotgun shell but there was no sign of the creature.[61]

Many people have reported seeing a Bigfoot-like creature in the peaceful hamlet of Lewiston on the Canadian border just north of Niagara Falls in extreme northwestern New York. Based on interviews with over four dozen witnesses by Lewiston village police officer Peter Filicetti, the creature stands about six feet tall and weighs 325 pounds, making it a foot shorter and perhaps 100 pounds lighter than reports in the Adirondacks.

Filicetti began collecting reports after his own close encounter while visiting his parent's farm in the autumn of 1976. On that September day, he said, "We were picking corn when I looked up and saw rows of corn separating."[62] He then heard a grunting noise and the sound of a large animal running. Upon inspecting the damage to the corn, he was stunned by what he found: row upon row of trampled corn and what appeared to be human footprints leading off into a marsh. He was adamant about the tracks which trailed for at least 200 yards: "They were footprints, not paw prints!" The strangest aspect of the incident was the prints, which had only three toes. Filicetti's police instincts kicked in and he quickly organized a hunting party with nine others, complete with tracking dogs, but he noted that "even with the dogs we couldn't find a trace of it."[63]

The Stump that Stood Up and Walked Away

In March 1977, Royal Bennett and his fourteen-year-old granddaughter Shannon were spotting deer from the back porch of his house on the west side of South Bay near Whitehall. As each focused their binoculars on an object that was several hundred yards off in a clearing near Fish Hill Road, they spotted what first appeared to be a stump. Suddenly, she said, it stood and walked off. It was amber in color and moved slowly. Shannon continued: "It looked like it was a little hunched over when it moved and light tan in color on the main part of the body. It was about one in the afternoon when we saw it. We couldn't see the face clearly, it was just too far away. I would estimate the height between seven and eight feet tall and it had long arms that swung as it walked. Then it entered the trees near the clearing."[64]

Royal, an avid outdoorsman who is at home in the wilderness, was convinced that they had witnessed something extraordinary. "If it was a man, it was an awfully big man…it looked honey-colored. Even from this distance with the aid of binoculars, I could tell it would weigh close to 500 pounds and it was tall…It's close to 100 feet across that clearing. No bear can walk that far on its hind legs, and...it was the wrong color for a bear."[65]

A year and a half later, in September 1978, two pheasant hunters in the town of Porter near Lewiston — Kevin Mooradian and David Holt—came across what appeared to be the body of a decomposing Bigfoot. The discovery caused a brief media stir and made headlines in the tabloids. *The Weekly World News* proclaimed: "Bigfoot Photos Baffle Experts…Family of Bizarre Creatures may be Roaming the Wilds."[66] While the skull disappeared — perhaps by wild animals, officer Filicetti snapped off a series of high quality photos of it. Despite the media speculation about the possibility of a dead Bigfoot being found, which appeared even in the mainstream media, when press photos of the creature were sent to two experts on mammals, they both came to the same conclusion. Dr. Sydney Anderson of the American Museum of Natural History in New York City, and Dr. Richard W. Thorington, Jr., a mammal expert at the Smithsonian Institution, were in agreement: the animal in the photos was a black bear.[67] The hunters who made the find may have jumped to the conclusion that it was Bigfoot, owing to the many stories of the creature being seen in the Lewiston region over the years. One hotspot is the Tuscarora Indian Reservation where Chief Frank Wasko says his people have heard stories about such a creature on their land for years.[68]

Investigator Bill Brann heard about a fascinating incident from the late 1970s, after interviewing two Saratoga County Sheriff's

Deputies who wish to remain anonymous. One summer evening, the first deputy was called to a reported disturbance and as he pulled in, noticed several people milling around a house. When he asked what was going on, they said that something was making loud screams behind a mobile home, then pulled up a tree and hurled it against the trailer. The officer went around back. Indeed there was the tree, some ten inches in diameter, leaning against the trailer. He found nothing else but brush that had been matted down by something big and powerful. He advised them to keep their doors and windows locked. The back-up deputy told Brann, "I know it happened. I listened to the entire conversation."[69]

The Kinderhook Creature

Kinderhook and its environs, fifteen miles south of Albany, has been the site of several sightings of what locals have dubbed The Kinderhook Creature. Nestled in a rural section of the Hudson Valley in northern Columbia County, the area has a reputation for strange events. Since the late 1970s, a series of encounters with a huge, hairy man-beast have been recorded.

In December 1978, a seventy-two-year-old Kinderhook grandmother named Martha Hallenbeck spotted a "big black hairy thing all curled up" on her lawn. Later she found huge, human-like tracks in the snow. She kept the incident to herself fearing that people would think she was "nutty." She said the creature looked tall and had long hair.[70]

On a summer morning in 1979, a witness reported that while hiking somewhere in a

Barry Knights' drawing of the creature he saw in Kinderhook. Note that he shows only three toes.

section of remote Hamilton County, what may have been a juvenile Bigfoot passed nearby. Becoming separated from his fellow hikers and moving off the main trail so as to observe a beaver colony, he heard a rustling noise in the thick brush. "I could hear a loud thump-thump of feet pounding the main trail and it was wailing like a womans (sic) scream…It ran partially into view and was about four and a half feet tall with dark brown fur with lighter tips and was moving fast upright."[71]

Martha Hallenbeck

On the afternoon of December 5, 1979, fourteen-year-old Barry Knights was trapping on Cushings Hill in Kinderhook, when he observed four huge furry creatures walking on two legs and crossing a creek. He said they had light brown fur and made clacking or grunting sounds. Martha Hallenbeck recounts the story: "My grandson Barry…came back…and he said to me, 'Grandma, I saw four great big things crossing the creek and going into the woods down there.' And I'm sure he saw something because I hadn't mentioned anything I'd seen before that. And then I told everyone that I had seen the tracks before." Knights was so spooked by the incident that he spent part of the day carrying a baseball bat for protection.

The next day, Albany radio personality and film producer Bruce Hallenbeck went to the scene of his cousin's sighting. Barry was

Bruce Hallenbeck (right) is seen here being interviewed by Doug Hajicek, television producer, for the Outdoor Life Network's series, *Mysterious Encounters: The Creature of Whitehall*, 2003.

still shaky from his encounter and refused to go with Bruce, so he took his other cousin Russell. While they found nothing, Hallenbeck said that the area around Cushing's Hill seemed like "another world." The area, which has a reputation for being haunted, is heavily wooded, covered in thick underbrush, and is pocked with bogs.

In April 1980, a woman was driving along Route 9 near Kinderhook when her car headlights shone upon a seven-and-a-half-foot-tall, reddish-brown creature that "looked like a highly evolved ape." She said it walked out of a field on her left, crossed the road and entered a wooded area to the right.

That summer, a Connecticut man got a shock during a camping trip in the rugged outdoors of Lawrenceburg in Upstate New York. Fred Renaudo, 34, said that on June 4, 1980, he was awakened by the cries of a coy dog or wolf — or so he thought. He sat up and scanned the woods but saw nothing. Suddenly he could hear the sound of something crashing through the brush.

"It was making loud thuds. I could hear it coming for quite a distance before I saw it, but finally I saw this white shape coming out of the darkness. It was walking at a pretty good pace, but the thing was that it was breathing real hard, like it had asthma, or it had been running real hard. There were big heavy noises...very loud...50 to 75 yards away."

When it came into view, Renaudo was startled by the sheer size of the creature, noting: "[T]his thing was big. It was large and white...and furry. I was looking at it pretty closely and even though I couldn't see it that clearly, I definitely had a feeling about what it was. Anyway, I got into the truck. I rolled the window down and stuck my head out and I could see it fading back into the darkness. And then it disappeared and I didn't hear it anymore."

Several days later he returned to the site with the property owner and found footprints fifteen inches long. Later he told a reporter: "I'd just assume they leave him alone. I know I believe it. I don't really care if anyone else believes it."[72]

On the bright, cloudless full-moon night of September 24, 1980, seventy-year-old Martha Hallenbeck of Kinderhook and several relatives, reported an encounter with a large "something" outside Martha's rural home soon after 11:00 p.m. As they were dropping off Martha after an evening with the family, they heard frightening screams in the woods near the house. Bruce Hallenbeck would later write in a letter, "She was terrified; it screamed, moaned, made guttural noises, and finally my nephew got his shotgun and fired into the air. It moved away, walking on TWO legs, such as a human would do." Bruce's letter was sent to popular *Albany Times-Union* newspaper columnist Barney Fowler, who wrote about it and published Bruce's request for other area residents to come forward with similar encounters. The column triggered a deluge of letters by area residents with similar accounts, including three near Kinderhook Creek a year earlier.[73]

Kinderhook was again the scene of Bigfoot activity one November night in 1980, when Barry Knights and Russell Zbierski were walking along a desolate road near Cushing's Hill. Suddenly, they heard the sound of something large moving in the woods on both sides of the road. What happened next left them stunned and shaken. Five hulking creatures with cone-shaped heads and no necks suddenly converged on the road ahead. They quickly fled in the opposite direction. At about the same time, a woman just down the road said she saw a huge hairy creature that walked on two legs, remove food from trash cans by her garage. Her dog was so frightened that it was spinning in circles wetting itself.

In early April 1981, a young woman was riding her bicycle on Novak Road in Kinderhook in the vicinity of many previous sightings, when she saw a huge creature cross the road and disappear into a cornfield.[74] About a month later on the night of May 8, several campers near Cushing's Hill spotted a tall figure with glowing red eyes. They said it walked on two legs, had no neck and long arms.

During summer of 1981, a family living in a remote part of Columbia County to the south of Kinderhook, reported encountering a black, hairy Bigfoot one night. The incident occurred just twenty minutes from Kinderhook at their secluded home at the end of a dead end road surrounded by woods. After showing the sixteen-inch long tracks to a wildlife expert from the State Department of Environmental Conservation, the officer responded, "If those are bear tracks, it's a real monster."[75]

One night in November 1981, Bruce Hallenbeck's cousin Chari was driving by her grandmother's driveway in Kinderhook when she spotted a "big two-legged thing, reddish-brown, that ran off into the woods." The creature fled as soon as her headlights shone onto it.

Another Remarkable Police Encounter

Three months later, two Whitehall police officers saw something extraordinary while making their rounds, traveling north on Route 22, half a mile from East Bay at the base of Lake Champlain. It was early February; the time was 4:30 a.m. Suddenly, they watched in disbelief as a tall hairy creature dashed across the road and ambled up a steep embankment at the base of a mountain. The incident occurred just 100 yards from the Washington County Highway Department Garage. The officers kept their identities secret, fearing ridicule.

In 2005, one of the officers decided to go public. His name is Danny Gordon, a well-respected, longtime resident. He said the creature had narrow shoulders, was lanky, and stood between seven-and-a-half and eight feet tall and was covered with dirty, mangy, dark-brown fur. Gordon was adamant that it wasn't a prankster. When pressed on the possibility that it may have been someone trying to give them a fright, Gordon responded, "There's no way in hell that I could believe this was a man in a fur suit." A more pertinent question comes to mind: who in their right mind would dress up in a monkey suit on stilts on a frigid winter morning and dash in front of a police car knowing that there were two armed officers inside? And how could it have maneuvered up the side of the mountain so easily?

Gordon said it looked like an ape with poor posture, as it slouched. Its arms were long and swung back and forth as it took huge strides. The speed at which it moved was remarkable, and he said "a relay runner would have trouble keeping up with [it]."[76] Gordon got out of the car and started tracking after the creature, with his gun drawn. Meanwhile, the second officer said he had no motivation to leave the car and was content to watch the proceedings from afar.[77]

Former Whitehall police officer Danny Gordon

Late one afternoon in May 1982, Michael Maab was fishing by a dam near Kinderhook when he got a feeling that he was being watched. As he looked across Kinderhook Creek he spotted an eight-foot-tall creature just twenty yards away that seemed to be looking him over. It was covered with short reddish-brown hair except for the head, which had longer hair. He was close enough to note that its eyes were small and beady, and it had black fingernails. The two stared at one another for two minutes before it ambled off into the woods.[78]

During the summer of 1982, an elderly Kinderhook resident told local investigator Bruce Hallenbeck that he saw a big, black hairy creature one evening at dusk. He said it was "standing out under a big tree in the yard." He was adamant that it stood on two legs. In September of the following year, another elderly Kinderhook man was traveling on Novak Road at 9:00 a.m. when a large animal walked in front of his car. It resembled a large black bear. He too was certain it was walking on two legs.[79]

A bicycle trail in the popular tourist resort of Lake George in Warren County, about sixty miles north of Kinderhook, was the site of a Bigfoot encounter on the evening of October 7, 1983. At 7:45 p.m., three males, ages fifteen to twenty-seven, were biking four miles south of Lake George Village not far from French Mountain, when they heard what sounded like screaming from a nearby hilltop. When one of the trio shone a light on the area, he saw the reflection of big red eyes seven feet from the ground. They quickly pedaled back to Glens Falls. They told investigator Bill Brann that when they returned to the site the next day, "there was a strong stench still in the air, and the smell did not resemble that of a skunk."[80]

Hunters often have strong opinions on the subject of Bigfoot. Many proclaim that in years of hunting, they've never encountered anything unusual. Others have had that notion shattered with just one experience.

A man reported that in November 1983, he was deer hunting in the Adirondacks when he said he came across Bigfoot near Spire Falls Mountain, in Palmer Town Ridge. "I can picture it like it happened yesterday," he said. "I was deer hunting...it was an overcast afternoon and I was sitting there...[near] the base of this mountain near the ridge. I hear this deer coming to the right and all of a sudden I see this person I thought was a guy jogging in a suit. It looked like he was wearing a fur coat…I looked again and I sat and said, 'Holy Jesus, what the hell was that?'" In talking with Paul Bartholomew, the man said the creature he saw looked just like the sketch made by Eric Miner of the Abair Road incident. Stressing his sincerity, "I'm not kidding you, I'd swear on my kid's life," he said.

The man said it appeared that the creature "was playing games with a deer" for twenty minutes. Every time this nearby deer would take a few steps and stop, the creature would look back over its shoulder at the deer. The man was able to get a good look at the creature with his binoculars, though he said he had a strong urge not to look at it.

"Whatever the hell it was, it was crouched over…slouched over…I was figuring maybe six feet at the most… it was hard to tell the way it was hunched over, but I noticed the arms were hanging down…and it looked like he had a fur coat on everywhere and had long hair." The man was adamant that it couldn't have been someone in a suit. "There's no man, no nothing that could run up the side of a ridge like that…the way it was moving."[81]

Between August 15 and 17, 1984, strange animal noises were heard by a rural Whitehall family between 10:30 p.m. and 12:30 a.m. One member said it was like an animal experiencing agony as if it were caught in a trap. The vocalizations endured about a minute. Then on the night of the 20th, one of the family members spotted a seven- to eight-foot-tall creature just twenty yards away. The time was 10:45 p.m. He turned to get his friend's attention but by the time he turned back, it was gone. He was certain it stood on two legs. Eight days later, the same witness spotted a similar creature near the same spot, walking by a house at 8:30 p.m. An outdoor light enabled him to get a good look at the figure which he said weighed "at least 400 pounds." After watching for ten seconds, it ran off.[82]

In Our Headlights

During the same month, a strange creature was sighted in the Capital District seventy miles to the south. An Averill Park couple was driving on a dark, desolate road just outside of Postenkill when they spotted a huge Bigfoot that was running. In a letter to Paul Bartholomew they said, right after rounding a bend in the road,

"...[W]e both observed a large upright figure vault a pasture fence approx. 100 ft. ahead of our vehicle. This figure was tall and reddish blonde in color and running very quickly. It was in the path of my headlights..." After turning around and scanning the area, he said "we could still make out a figure running upright downhill across a pasture...I had no immediate logical explanation for what we had just witnessed, and experienced a feeling of absolute fear and dumbfoundedness difficult to describe."[83]

Five years later, a rare daylight sighting was reported by a sixteen-year-old boy and his dog in August 1989. In a story for *The Whitehall Times*, Paul Bartholomew recounts the incident.

> A startled teenager may have recently glimpsed a creature resembling the legendary "Bigfoot"...in Hampton, New York.
>
> The 16-year old, wishing to remain anonymous, claims he was walking along 22A South when he spotted a dark figure lurking along the roadside. The sighting reportedly took place around 3 PM on August 18.
>
> The youth reportedly observed the creature for approximately three full seconds before both took off in opposite directions.
>
> "I could see the back, a big back, and it was dark, it was awful matted and scruffy-like," said the witness.
>
> The youth was accompanied by a dog which apparently reacted to the presence of something.
>
> "The dog...came across the road near where the thing was standing and it started whining and took off," said the teen.
>
> On August 24, the mystery deepened as the young man and a friend camped out near where the sighting took place. At 2 AM it seemed that something was watching them.
>
> "I stared at it and it stared back and it kind of went away," the youth said.
>
> The witness said he observed two "glowing red eyes" approximately six feet off the ground. The eyes appeared to be circling the camp.
>
> "They just weren't round like, they were oblong, longer than they were round," added the observer.[84]

An examination of the campsite found two footprints corresponding to where the glowing eyes were spotted. They were ten inches long and four inches wide. The distance between them: fifty-six inches. This second encounter took place near where thirteen

Possible footprint found at Hampton, New York during August 1989.

years earlier several witnesses, including State and Whitehall police, reported seeing a large, hairy creature with glowing red eyes, that disappeared into woods.[85]

In January 1990, footprints measuring twenty inches long were found in a field at Ghent, New York, trailing through the snow. Several days later similar tracks were found, leading into a thicket, after which they disappeared.[86]

In the autumn of 1990, a man observed a creature "on two legs" as it ran down a bank in a wooded area along Hickey Road in rural Whitehall, New York. The man said, "A bear would get up to look around, smell, but this here, it was dark and it went down over the bank." The man was resolute: "I knew it wasn't a bear."[87]

A Bigfoot sighting can be an experience that will last a lifetime. In the autumn of 1993, while waiting for the school bus near Abair Road, a brother and sister, ages twelve and thirteen respectfully, saw a Bigfoot-like creature that frightened them. They said the figure was hunched over and looked like it was eating. Both siblings were shaken by the sighting even though it lasted but a few seconds. The sister was especially upset by the encounter.[88]

Halloween is always connected with strange creatures and scary tales. In 1994, one resident of Abair Road in rural Whitehall encountered a mysterious creature befitting the day. At 7:10 a.m., the man spotted a dark, hairy, brown mass "walking" away from about 400 yards. "It was either that (Bigfoot), or a big moose," he said. He went to the area and saw what appeared to be matted grass where something had been. The grass had a lot of dew on it, nothing firm.

"I did see it move. It took one step and it was out of sight" (unlike a moose movement), he said. He described it as "big and huge." There had been a lot of logging going on recently on both sides of the site. The man added that whatever it was appeared to be bent over — about four feet off the ground.[89]

On March 14, 1995, Darrin Gosselin and a friend were hiking

off Pike Brook Road in rural Clemons, bordering the town of Whitehall. The pair came across a long trail of tracks in the snow. While the track trail was subject to spring snow melt, the stride between the tracks measured only four steps to his six steps. Both men said the tracks were pretty long and at one point the track-maker jumped from a ledge-like area. There is a large, abandoned graphite cave nearby, with good ventilation and reaches at least one hundred yards in length, from five to ten feet in height. Paul Bartholomew and Darrin Gosselin walked the entire length of the cave and believe it would certainly be adequate to support a sizeable creature.

A Possible Baby Bigfoot

A strange squatting creature was spotted by two workers in early June 1995 along Route 28, coming out of the north creek and heading towards Olmsteadville, a tiny, rustic hamlet south of Minerva in southwestern Essex County. While driving along at 12:25 a.m., the pair was startled by a five-foot figure on the side of the road. They said it was very thin, weighing about 80 pounds.

The driver said: "It was about the size of a ten-year-old boy. Its arms were long and resting on its knees. It looked as if it had brown and white matted hair. We did not see a tail. It had a human-shaped head and shoulders. It looked right at us and the eyes didn't shine in the headlights like other animals do. The eyes were kind of big and oval shaped. It seemed unafraid, but yet it had a gentle look about it. It didn't startle and run. It stayed there in the same position."[90]

On August 15, 1996, two people were fishing from their canoe at dusk at Pine Pond near Saranac Lake when one of the witnesses noticed a mysterious figure fifty yards away at the edge of a wood line. At first it appeared to be a black bear, as they estimated it was three and a half feet from the ground. The pair began to paddle cautiously towards the figure. When they got to within forty feet, it suddenly stood up, startling them. One of the witnesses said it stood seven feet tall and had dark brown hair. "Its face was hairy yet fleshy around the upper cheeks. Its eyes were dark in color but clearly visible and had a brightness about them." The creature stared at the pair for ten seconds, before tilting its head as if to sniff the air. At this point the sound of snapping twigs could be heard behind the creature. It turned its head to one side, then "immediately turned back towards us and then spun 180 degrees around and darted into the woods like a cat. They could hear the creature moving through the woods for about ten seconds.[91]

The incident left one of the pair shaken. "The whole experience was very, very upsetting. Although I can honestly say it did not attempt to threaten us…it was scary as hell. That night I did not sleep one wink." The encounter took place at 8:00 p.m. on a beautiful summer's day.

The tiny hamlet of Modena in the Ulster County town of Plattekill in the Catskill Mountains, was the scene of an unusual Bigfoot claim in 1997. A Lembo Lake camper, Doug Pridgen, was videotaping during a rock concert and noticed nothing unusual. However, later while viewing the tape, what appeared to be two ape-like creatures were discerned in the background 100 feet away. The taping took place in an old farm orchard. Pridgen said, "At first, you think, 'big monkey,' but what would a monkey be doing in the woods of N.Y.? The thing swings from one tree to the next and moves up and down it with ease."[92]

A young woman recalls that as a twelve-year-old girl in February 1998, she accompanied her cousin for a walk near the Genesee River near the cousin's home. The girls knew the path well. They soon heard rustling in the woods, that they presumed to be a deer. The noises persisted for three minutes. Half a mile down the path they came to a clearing with swampland on one side and woods on the other. Suddenly, a tall, grayish-white creature jumped onto the trail in front of them, coming within two yards of the pair. The creature stood six-and-a-half feet tall, had broad shoulders and a muscular physique. Being so close, she got a good look at the body. She said the hair was three inches long and its arms swung as it moved. The hair was thick, shaggy, and matted in places. It made a growling noise. The creature walked quickly and was visible for about thirty seconds, after which time the girls ran off.[93] The encounter took place near River Road two miles from the town of Wellsville in Allegany County.

It Stood Just Twenty Feet Away

In late August 1998, a Bigfoot creature was spotted on a remote dead-end road near Caroga Lake in Fulton County on the southern fringe of the Adirondacks. Two men were riding on North Bush

Road at 2:00 a.m., on a chilly, starlit morning, when they pulled their truck to the side of the road to urinate. One of the witnesses, Chris, returned to the vehicle before his friend, and he caught a glimpse of a huge figure as he flicked on the high-beams of his headlights, standing twenty feet away. It was seven to eight feet tall, part man, part animal, and had long, brown hair. Chris said, "It stood perfectly still for a minute and then grunted at us. Then it turned and walked away. It didn't move like a man. It kind of swaggered back and forth like it lunged each leg forward when it walked." He was adamant that it clearly walked on two legs. He said the face was flat, and the arms swung in an exaggerated fashion.[94]

On September 13, 1999, Bigfoot was sighted in the Finger Lakes region of the state. A man said he was driving at night along highway 224 near Corning in Schuyler County, when the lights of an oncoming vehicle illuminated the creature leaving a cornfield and heading towards a wooded mountainous area.[95,96]

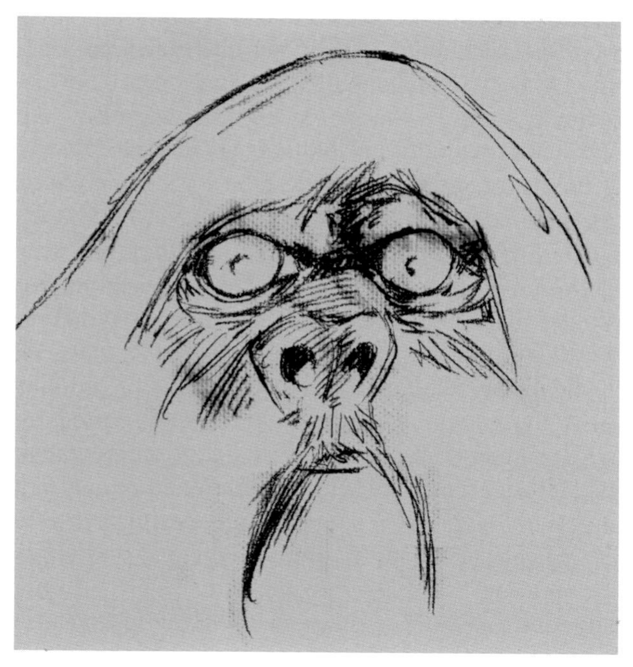

Eyewitness descriptions of the creature often include glowing red eyes and pig-like snout. Drawing by Eric Miner.

3 21st Century Reports

> I have been an avid outdoorsman since a very young age, hunting, hiking and camping for over forty years and very experienced with animals in the wild and I know what I saw that afternoon was remarkable.
>
> —Anonymous witness [97]

Encounter at a Girls' Camp

During the summer of 2000, a group of up to ten young female hikers got the shock of their lives while walking the Moose River Camp Trail near Lyon Falls in Lewis County on the northwestern fringe of the Adirondacks. On the afternoon of July 11 while hiking near their summer camp, a creature was spotted in a clearing 100 yards away. One girl noted, "We came into a swamp-like area and heard some twigs snap and smelled this awful smell and saw something moving in the trees ahead. The creature was dark, hairy, gorilla-like, and it was walking on two legs..." She said it was in view for about ninety seconds, and when one of the girls screamed, it stopped and looked in their direction. The group fled back to camp, where she said the counselor suggested that they had seen a bear; she was certain it wasn't, pointing out that it walked on two legs for over a minute. She thought the counselor may have been trying to prevent a panic at the camp.[98]

The Alleghany Reservation near Salamanca in Cattaraugus County, was the site of a Bigfoot encounter reported by a local hunter in the autumn of 2001. The incident happened near dusk along Old 17 Road when the witness heard something large crashing through the woods, then spotted a tall creature with dark hair, walk over the railroad tracks. A skunk-like smell was noticed at the time of the sighting.[99]

In the autumn of 2002, a hiker and avid outdoorsman was driving near Middleport in Niagara County, when he pulled into a remote driveway in hopes of spotting deer. The time was 6:00 p.m. As soon as he got out of his vehicle, he noticed a dark, two-legged creature moving quickly in the field about 225 yards away. Its hair was long, dark brown and flowing as it moved with its arms by its side as it almost seemed to float across the rocky, uneven terrain and tall grass. "When I say traveled very quickly, I mean it was really moving very rapidly but at no time did it seem like it was running as you or I would look if traveling at that speed...[as] there was no discernable movement of its arms as it 'floated' through the tall grass." The encounter made a deep impression on the man who said he was convinced it was a Bigfoot.[100]

In June 2003, Larry Paap of Granville, reported seeing a strange creature squatting in a field off Route 22 near Comstock's Great Meadow Prison. The tiny hamlet of Comstock borders the Bigfoot hotspot of Whitehall. "I slowed down and got a good look dead on thirty yards away," he said. Staring at the squatting figure he saw a head and wide shoulders. The figure was said to have "golden-brown hair." He said the hair was four to six inches long and really well kept and "beautiful" as it blew in the breeze. "It had a clean look to it." Paap said the head had a curious shape, but was like "a cross between a dog and a cat." He said the head seemed to be "thicker." He could not discern any ears or whiskers. He said the creature was four feet off the ground as it crouched over and had massive shoulder blades and its legs appeared as though they were straight out on the ground "as if it was sitting." The neck was "stretched six to eight inches." Paap said he and the creature stared at each other for at least half a minute, the figure remaining motionless for the entire time.[101] The next minute, Paap said, the creature seemingly vanished before his eyes. "It was like it disappeared into thin air," he said.[102] He told reporter Patrick Ripley of *The Whitehall Times*, "It just blew my mind. I had never seen anything like this before." He also said, "I got a really good look at it. I was pretty scared...it was just phenomenal...I am still kind of in awe."[103]

A Giant Flat-Faced Monkey

One morning in June of 2004, two visitors from Hong Kong were spending a leisurely day fishing just north of Whitehall in the rural, mountainous town of Clemons. Their fishing spot was in back of the Redtop Tavern, about ten miles north of the Village of Whitehall. That's when they spotted a large monkey-like creature wading through the water. One of the men said it was tall and skinny and stood up in the water sixty yards away. He said its body stood at least five feet out of the water and was clearly visible from the waist up. He described what he saw as a monkey. He told Paul Bartholomew: "I remember...noise and sound (dogs barking)...then looked and (saw the creature) in the water up to its chest...then it moved very fast," he said, and quickly disappeared out of view. Surprisingly, not being familiar with North American wildlife, the men didn't realize that the encounter was all that unusual and went about their fishing, oblivious to the history of Bigfoot sightings in the area. One of the men said the creature resembled an orangutan, with short, brown and red hair. "It had a flat face," he added.[104]

On June 20, 2004, three men spotted a Bigfoot-like creature in Monroe County near the town of Greece. The witnesses, ages 45, 43 and 36, were on a bike ride with two companions at 12:30 a.m.—a ride they made about three times a week along a local bike trail. While taking a rest, they were standing around talking when their attention was drawn to a figure standing by a tree about seventy-five yards away. As they walked toward the figure to get a better look, it ran off down a wood line about 100 yards. They said it stood seven and a half feet tall, was covered in tan or gray fur, and ambled off into the woods on two legs. The sighting area was near the southeastern shore of Lake Ontario, and situated near dense woods and wetlands.[105]

On October 15, 2004, a father and son turkey-hunting duo were in a hilly, forested section of Napoli in Cattaraugus County in Western New York, when the son spotted a strange creature at close range. The man, age twenty-nine, said he had slightly separated from his father when he could see movement ahead. Readying his weapon and expecting to see a turkey, he was suddenly startled when he saw the creature and couldn't bring himself to shoot. He said the thing was just twenty yards away. "I didn't know what it was. It was on two legs, all brown-red hair all over it and at least six feet tall." Despite being stocky, it ran swiftly down a hill and out of sight after five seconds. His father was looking in the wrong direction and saw nothing. The encounter happened at 11:00 a.m. at the top of a hill.[106]

Late on the morning of June 10, 2006, a motorist reported seeing a Bigfoot-like creature while driving near Minerva in Essex County. The witness had traveled seven miles along the unpaved

North Woods Club Road and was driving fast when he and his passenger, a brother, reached a bend in the road. "As we turned a bend I saw a very large, brown, furry object lunge into the dense woods up that road at the next curve fifty yards ahead. It was obvious that the creature was in the road and was startled by my speeding SUV." As they approached, they saw what appeared to be the head and shoulders of a large creature that stood seven-and-a-half feet tall. Its hair appeared to be "a light chestnut brown." The witness was certain that it wasn't a bear, and was surprised that it vanished so quickly. The thing was in sight for about three seconds.[107]

On Sunday night September 3, 2006, at 9:40 p.m., a family was traveling through Whitehall from nearby Rutland, Vermont on their way home to Glens Falls, when they reported seeing a Bigfoot. The driver, Rich Martin, said the figure was crossing Route 4. It stood six to seven feet tall and was "covered in dark brown hair all over."[108]

The next day, four people were driving along Route 4 into Whitehall at 8:45 p.m., returning from the Rutland State Fair, twenty-six miles to the east. As they passed by a series of storage shed rentals near Whitehall High School, they spotted a mysterious figure standing beside the road. "It had a white face and black hair," said one witness, and stood six to seven feet tall. The man who first spotted it said he waited several seconds before saying aloud, "Did you guys see that?" As it turns out, the driver and all three passengers glimpsed the creature. They all gave corresponding accounts and drew similar drawings of what they had seen. The driver described it as standing "tall, with two dark eyes close together." There was no description of red eyes—a common feature of Bigfoot reports. The sighting lasted three to four seconds, but was enough to make a lasting impression. When asked if she could have been mistaken, she replied, "I definitely saw what I saw."[109]

The next month three teenaged girls camping in rural East Whitehall, got the fright of their lives while walking down a footpath some 200 feet from their campfire. The encounter took place a mile east of the Old Brick Church at 10:30 p.m., when they spotted a large figure thirty feet away. "It was like a human," said one girl. "We all saw it at the same time and we all screamed." Two of the trio immediately ran back towards the makeshift camp, while a third girl just "froze." "That thing was just standing there," said the seventeen-year-old. By the time she gathered her thoughts and looked around, her friends were gone. She then turned back to where the creature was standing, and it was gone.

She said the creature "was huge — seven feet," formed a "big outline" and petrified her. She estimated that the sighting lasted eight seconds. "It was big — wide, and looked like a human shape or form," she added. The sighting occurred near a rural shooting range. Curiously, two or three weeks earlier, the girl's family said their dog was visibly upset as they heard what sounded like "steps stomping" around outside. There was no odor associated with the sighting.[110]

PART 2 BIGFOOT Encounters in New England

4 Historical Sightings

> Suddenly he raised his giant arms above his head and waved them several times... turned and continued along the ridge with the agility of a gymnast. ... this was no ordinary man.
>
> —ELDERLY MAN DRIVING NEAR TINMOUTH, VERMONT

Nestled in the northeastern corner of the country, the New England states of Vermont, New Hampshire, Maine, Connecticut, Massachusetts and Rhode Island encompass a vast expanse — 69,746 square miles. Outside of the major metropolitan areas of Boston and Providence, much of the region consists of small towns which dot the landscape among fertile valleys and rugged mountains that were carved out by receding glaciers. Much of the northern region is heavily forested with maple, hemlock, birch, beech and pine trees. New England is home to the White Mountains of New Hampshire, the Green Mountains of Vermont, and the Berkshire Hills in western Massachusetts. Maine alone has 17 million acres of forests.

The earliest recorded evidence of a Bigfoot-like creature inhabiting New England can be found in the novel *Northwest Passage* by master storyteller Kenneth Lewis Roberts. It is one of the greatest pieces of historical fiction ever written. In the book, a scout for Major Robert Rogers and his famous band of rangers tells of a 1759 encounter with a strange "black bear" during the French and Indian War. This incident is the first written record of a Bigfoot-like creature in what is now the state of Vermont. The "bear" reportedly tossed pine cones and nuts at the party from afar while they were camping in northern Vermont near Missisquoi Bay, the section of the lake that juts six miles into Canada. According to a scout named Duluth, the Rangers sailed from their camp at Crown Point, up to the northern tip of Lake Champlain on the Canadian side of the lake. There, they attacked an Abenaki Indian encampment at St. Francis in what is now Quebec Province, as payback for an earlier attack on a group of British soldiers. Duluth wrote that the "bear" encounter occurred while the Rangers were making their getaway; the group was being hotly pursued by a band of Indians and by French troops. Duluth said that his raiding party [in his own words] "were ever being annoied, for nought reason, by a large black bear, who would throw large pine cones and nuts down upon us from trees and leges."[111] He said the Native Americans were "disgusted" by the creature's acts of harassment, "and knowe him, and call him Wejuk, or Wet Skine."[112]

Pioneer Incidents and the Legend of Slippryskin

During the latter part of the eighteenth century, the first settlers of northern Vermont heard tales of a Bigfoot-like creature bearing the name Slippry Skin and slippryskin. The name can be found in numerous oral and written traditions throughout the state and is suspiciously similar to the Indian name of Wet Skine that Rogers' Rangers reportedly encountered. Descriptions of a similar creature in the region at about the same time period would seem to be more than a coincidence. The nickname was given on account of its remarkable ability to elude hunters. Early Vermont historical records abound with encounters of Old Slippry Skin, which was said to resemble a huge bear. However, unlike any known bear, it mostly walked or ran upright on two legs. Slippryskin encounters fill the histories of several northeastern Vermont towns. No doubt many reports have been embellished, but if even partly true, it would be difficult to believe that the creature was an ordinary bear.

Historian Paul Rayno has uncovered numerous references to this Bigfoot-like creature in what is now Vermont. Rayno reports that during the latter decades of the eighteenth century, this strange beast was said to chase domesticated animals such as sheep and cows, and toss stones at schoolchildren and hunters. The animal was blamed for robbing meat from smokehouses, knocking over haystacks, placing stones in sap buckets, and wedging wire into mowing machines and hayrakes. Occasionally farmers would be shocked to find large rocks in or on their farm machinery.[113] Of course, it's entirely possible that rebellious Native Americans, disgruntled neighbors, mischievous youths, and even scorned lovers were responsible for many of these incidents.

The creature wreaked havoc in the Essex county towns of Lemington, Victory and Maidstone. In Lemington, it was credited with tearing up gardens, knocking over manure piles, and pulling down fences, leaving a trail of huge prints in the snow and mud. "Wet Skin" moved about in a ghostly fashion and was said to "vanish into the woods as silent and swift as a drift of smoke." Its cunning was said to be rivaled only by humans.[114] There are similar accounts from nearby Victory and Maidstone. In Victory, this "bear" was blamed for stampeding cattle and ruining cornfields by dragging small trees through them. It was said to be especially adept at throwing hunters off its trail. According to one account, the creature "would back right back in the tracks of his forward movement, with such accuracy and precision, that no one who did not suspect the trick would see any signs of a reverse movement, until the bear came to some large rock, or knoll, then give a long leap on to a bare spot, and move off in a direction diverging from that just pursued."[115]

One famous encounter with Slippryskin was reported to have involved then Vermont Governor Jonas Galusha, a renowned hunter. Galusha served two terms in the state's highest office, from 1809 to 1813, and again between 1815 and 1820. During one re-election campaign, Galusha said that he would rid the state of Slippryskin by personally shooting the creature. When it was spotted in Maidstone, he headed up a hunting party and traveled to the township within a few days, perhaps as a publicity stunt in order to boost his sagging campaign. Just before the hunt got underway, he is said to have pulled out a bottle of female bear scent and rubbed it on his clothes, hoping to lure Slippryskin out of the woods. He was unsuccessful — as was his re-election.[116]

Another party of hunters left the town of Morgan in Orleans County, with the intention of bringing the creature in. As they were walking along a logging road leading to the top of Elon Mountain, a loud thumping noise could be heard from above — the same sound that the Native Americans of the region attributed to the Stone Giants. The men quickly hid in some nearby bushes in expectation of ambushing Slippryskin. They returned home frightened and bewildered after claiming that the clever creature backtracked on his prints, then rolled a large tree down the mountainside, narrowly missing the hunters as they were lying in wait. They promptly gave up the chase and the hunt was abandoned.[117] How much is true and how much is embellishment and legend, is difficult to say. What is not in dispute is that a large Bigfoot-like creature with unbear-like qualities (such as walking on two legs, and human-like intelligence), was reportedly seen by many of the State's early settlers. It may sound incredulous that many hunters of the period reportedly took shots at the creature but failed to bring it down. Vermont Historian Michael Pluta observes that there may be a logical explanation for all the shooting failing to stop Bigfoot. "Typically, rifles of the period and region were light in caliber (.36–.45) or also low velocity. Penetration, typically, on heavy game would be shallow. Only persons who regularly hunted moose and bear would have spent money on a larger weapon."[118]

In August 1861, there were numerous sightings of an ape-man in the vicinity of Bennington in southwestern Vermont, and just over the border in North Adams, Massachusetts. According to a newspaper reporter, Vermont residents described the creature as "hideous." Several people shot at the man-beast, but it always managed to escape into the woods. There were rumors that it was a prankster wandering about the countryside in a gorilla outfit, as is so often the case with Bigfoot reports. But why would anyone run around the wilderness in a gorilla suit — on numerous occasions, no less— knowing that hunting parties were in hot pursuit, and continue to do so even after shots had been taken.[119]

A Mysterious Encounter by Hunters

On October 18, 1879, *The New York Times* published a front-page account of a strange creature sighted in the wilds of central Vermont.

> **A WILD MAN OF THE MOUNTAINS**
> **Two Young Vermont Hunters Terribly Scared**
>
> Pownal, Vt., Oct. 17—Much excitement prevailed among the sportsmen of this vicinity over the story that a wild man was seen

> on Friday last by two young men while hunting in the mountains south of Williamstown. The young men describe the creature as being about five feet high, resembling a man in form and movement, but covered all over with bright red hair, and having a long straggling beard, and with very wild eyes. When first seen the creature sprang from behind a rocky cliff, and started for the woods near by, when, mistaking it for a bear or other wild animal, one of the men fired, and, it is thought, wounded it, for with fierce cries of pain and rage, it turned on its assailants driving them before it at high speed. They lost their guns and ammunition in their flight and dared not return for fear of encountering the strange being.

A spate of wild-man sightings were recorded in Litchfield County Connecticut in the northwestern part of the state near the Litchfield Hills. The outbreak began in Winsted during August 1895. The sightings caused a sensation throughout New England as journalists soon descended on the community. The saga began on a humid summer day when local Selectman Riley Smith said he spotted the creature. It was August 17 when Smith said he stopped to pick blackberries along the old Colebrook Highway near Old Losow Road five miles outside of Winsted in an area known as Indian Meadow or Injun Meadow. He said the creature emerged from the middle of the berry patch. Its body was naked and "covered with hair." Its head was much bigger than a human's, its hands were exceptionally large, and it had sharp teeth that were the size of a horse. When the creature appeared, he said his bulldog Ned began to whine, "and with its tail between its legs, sought refuge in … [a] wagon under a pile of blankets."[120] Smith said after the man-beast emerged from the bushes, it gave off fearful cries before racing into the woods and disappearing from view. As it ran off, he said its long black hair was streaming back from its head.[121]

Smith was popular in the region and was known for his physical strength, but he told *The Winsted Herald* of August 21, that the encounter left him filled with fear.[122] *The Winsted Evening Citizen* of August 27, said Mr. Riley had a solid reputation, which only added to the claims. It said, "Mr. Riley Smith is a man that talks but little: He is a man of undoubted pluck and nerve, and his word is first class."[123] The sightings triggered a scare that culminated in 500 people taking part in various search parties that were formed to hunt down the creature.

Another group of eyewitnesses was headed by John G. Hall who operated a stagecoach between Winsted and Sandisfield, Massachusetts. As the coach passed through Colebrook, Connecticut, a large creature crossed the road and jumped over a stone wall. As the coach approached the animal, Hall said it stood erect on the highway ahead. Hall pulled out a revolver and halted the horses. As he took aim, it ran off on four feet into Indian Meadow.[124]

Two New York City residents who were spending the summer in Colebrook, also claimed to see the creature in Indian Meadow. Miss Sadie Woodhouse and a Mrs. Mushone said the animal stood upright, was six-and-a-half feet tall, with black hair, white teeth and a very muscular form. A popular theory at the time was that the creature was an escaped zoo gorilla, though no such animals had been reported missing, and no body was ever found.[125]

There were sightings of a similar creature during the previous winter in the nearby town of Norfolk, Connecticut. Mrs. George Marvin of South Norfolk said she watched one morning as it stole one of her hens, while a Sandisfield farmer said he saw the creature grab one of his rabbits. Two other area men, Carl Moore and Joe Bruley said they fired at the creature with bird shot but it had no effect.[126] Norfolk resident Charles Benson said the thing jumped out of a tree before chasing him into his house. The previous spring, the gorilla-like creature was spotted by several people in Norfolk, entering a hole in the side of a mountain. They reportedly strung heavy chains over the hole to prevent it from leaving. As the story goes, the next morning, the chains were broken.[127]

In early July 1909, police in the northeastern Massachusetts city of Haverhill, searched in vain in a wooded area near Gile Street for a mysterious wild man who was spotted by local residents. The woods extended to the border with Newton, New Hampshire.[128] During World War II, there were several sightings of a wild man in the vicinity of Pontoosuc Lake just east of the Pittsfield State Forest in western Massachusetts. The mystery was apparently solved on November 18, 1942 when Massachusetts State Police arrested Willis L. Brazee of Pittsfield. He was a Private in the Army who had been reported absent without leave from Camp Edwards at Falmouth.[129]

5 Contemporary Cases

We never saw anything like it before. It wasn't a bear.

—GERARD ST. LOUIS

Modern-day Bigfoot sightings in New England began on a cold February morning in 1951, when lumberjack John Rowell was with a colleague named Kennedy snaking logs from Sudbury swamp south of Middlebury in east central Vermont. They noticed something strange. The night before, they had placed a heavy oil drum on a tractor seat and covered it with canvas. Returning the next day, they couldn't believe their eyes — the drum had been carried several hundred feet toward the woods. Around the tractor were giant footprints. The drum weighed 450 pounds. Rowell got his Polaroid camera and took several pictures, then gave them to a Middlebury newspaper in hopes that they could be published one day to publicize the incident, but they never appeared. The tracks were twenty inches long and eight inches at the toes. Curiously, the toes were turned straight down, as if from the tremendous weight of carrying the drum.[130]

During the early 1960s, Plainfield, Vermont farmer William Lyford heard his cows bellowing at Lanesboro Station along the Wells River–Barre Railroad line. Checking to see what the fuss was about, he spotted a tall bear-like creature walking on two legs. As soon as he shone his flashlight on it, the figure turned and ran over a hill, vanishing into the darkness.[131]

Eight Feet Tall and Crossed the Road in a Single Bound

One evening in 1964, Stockbridge in central Vermont was the scene of a spectacular sighting by six people. The witnesses were riding in a pick-up truck at 7:30 p.m. on a lonely stretch of road when a hulking grayish creature crossed in front of them. John Rose of Castleton Corners said the figure stood between seven and eight feet tall and strode briskly. Rose said the creature came out of some bushes and crossed the road fifteen to twenty feet away. He said it was definitely walking on two legs and crossed in a single stride, "like when you're going to cross a little stream...you push off and you land with your foot...out in front of you...just like one stride."[132]

In November 1968, the Newton Farm in South Kent, Litchfield County Connecticut was the site of a reported encounter with a huge creature. The witness said that when he was seven years old, he was playing in his house when he happened to look out a window and saw a hairy creature that was nine feet tall. He said, "I stood watching it walk along when it appeared to look up and make eye contact with me for only a second. It never broke its stride. Its eyes glowed like a raccoon or a cat at night when light is shined (sic) in their eyes."[133] The man now works in the concrete business and still vividly remembers the encounter. He says the eyes seemed to be self-luminescent and emitted an eerie glow. The sighting lasted five to six seconds.

In about 1969, a deer hunter says that in mid-November he came within 100 yards of a Bigfoot-like creature while tracking deer on a snowless mountaintop in Southern Vermont. Peering through a 9x rifle scope for thirty seconds, he said most of its hair was dark brown, though some was "frosted." The face had little hair, and its body was very large. He said it was holding something whitish — perhaps a bone or stick — and appeared to be tracking a wounded whitetail doe. "It stooped twice to smell the ground." After the creature had moved off, he walked to the spot and found large impressions in the ground near the creek bed, obviously made by something big and heavy.[134]

The mid-1970s were marked by several sightings in central Vermont. During mid-March 1974, two men were driving in Barre when they pulled to the side of Country Club Road to urinate. Suddenly there was a loud shriek. They outlined what happened in a letter to Castleton State College anthropologist Dr. Warren Cook who investigated Bigfoot reports in Vermont from the 1970s until his death in 1989. Cook was a respected scholar who was once nominated for a Pulitzer Prize.[135] The letter stated, "We both looked toward the direction from which the shriek came. We could see a tall, dark figure running across a field that had a dusting of snow...What amazed me was the speed at which this creature could run and at the length of its arms...the hands swung below the knees as it ran."

During May 1974, about seventy miles south of Barre, a farmer and his family living near Rutland, reported seeing an eight-foot-tall creature with black hair on their property. The hair on its chest was described as white, and it appeared to be digging up roots in a pasture.[136] That same year a couple said they spotted an eight- to ten-foot-tall creature in a field near Rutland. As police were investigating the incident, they saw a similar looking creature dash across the road.[137]

On a cool, exceptionally dark Sunday evening in September 1974, a student attending the University of Rhode Island had a terrifying encounter while riding her bicycle along Perry Avenue in the town of Wakefield, Rhode Island, five miles from the Great Swamp Area. Noticing that one of her pull brakes needed adjusting, she stopped near a street light. Suddenly she heard loud thuds from what sounded like crushing footsteps coming from a small field directly in front of her. A nearby dog began barking like crazy. Frightened, she began to slowly pedal away when she looked back and saw a huge, gorilla-like creature with white hair, step into the streetlight twenty-five feet away. The sight of the hulking creature terrified her. She said it stood six feet off the ground and weighed 400 pounds. It had "massive arms" and its knees were bent slightly. She emphasized that the shoulders were enormous. The head was "higher in the back" and was "connected to the neck all in one mass." The face was flat, dark, and it had deep set eyes that were close together, while its nose and mouth looked human. Strips of black hair ran down each side of its mouth. She said the creature got "an explosive start" as it began to chase after her. "My heart was pounding out of my chest, my eyes had tears in them and with all my might I pedaled as it ran on two legs, then down on its knuckles, then back up again…I remember how its long hair on its arms moved with each reach…" After about five seconds, the creature stopped and was "swaying back and forth" in the street, then turned and jumped a rock wall "all in one motion from a grass sidewalk."[138]

Extreme northwestern Vermont was the backdrop of a sighting just south of the Canadian border between August and early September 1975. A man was resting on a hilltop with his brother and son near Missisquoi Bay on Lake Champlain, when they saw a huge figure eight feet tall lumbering near the woods a mile and a half away [binoculars assumed]. The Bay extends several miles into

Canada and represents the northern most tip of the lake. The incident happened on the Vermont side of the Bay near the spot where Rogers' Rangers reported their 1759 encounter with a mysterious "black bear" that pelted them with pine cones and nuts. One of the witnesses described the encounter in a letter to Dr. Warren Cook: "I distinctly recall my brother exclaiming 'What the hell was that?' We all watched the creature for a period of several minutes, walking at a casual pace along a tractor-trail…It was dark in color…long in the legs. Wide in the shoulders. Comfortably upright…gait leisurely. It seemed to be in slow motion and yet it covered a large distance in a few minutes." The creature then turned and vanished into the woods.

The following year, remote Sawyers Mountain in East Haven, Caledonia County Vermont was the scene of a reported encounter. A woman said that she was sitting on the ground reading near an apple orchard off Route 114 at noon when she got the distinct impression she was being watched. Looking around, she spotted a large hairy creature that stood perhaps nine-and-a-half feet tall, quickly moving towards her. It came to within twenty feet, before moving off. It had a large head and muscular shoulders. The figure was said to have ape-like hair, "long arms and appeared to have human-like eyes." Curiously, the creature made no noise as it passed by, seemingly as if it were in a hurry — "walking very gracefully, as if to glide." While feeling uneasy but not threatened by the hulking figure, she decided she had better walk home, where she reported the incident to local police.[139]

Shortly after Christmas in 1976, Agawam in southwestern Massachusetts, gained national media attention after twenty-seven-inch, human-like footprints were discovered in a patch of woods off Moore Street (near Robinson State Park). Shortly after the find on Monday evening December 27, the press got wind of the discovery, drawing several Bigfoot investigators to search the locality and camp out by the tracks. During the week dozens of local residents kept visiting the site.[140] In early January, the tracks were revealed to have been a hoax after an Agawam teenager came forward to police and admitted making them. Sixteen-year-old David Deschenes told police that it took him two days to make the Bigfoot prints from boards. He then strapped them to his feet and trekked along the Westfield River on December 20. He said, "I did it as a joke for the little kids around here, but it got out of hand. The next thing I knew the police were out at two in the morning looking around, taking it seriously." As for why he didn't confess right away, he said, "I didn't feel like going out to tell them I was Bigfoot."[141]

In March 1977, a Chittenden, Vermont housewife glanced out her picture window and received a fright — a huge, hairy ape-like creature was standing in a field. The woman, who wants to remain anonymous, was certain it was not a bear standing on its hind legs. Its shoulders were slouched as if it had poor posture, and its arms hung at its sides. She said it was staring at the horizon. Frantic, she grabbed the telephone and called her husband who was nearby, but by the time he arrived, it was gone.[142]

Saturday May 7, 1977, was a memorable night for police in Hollis, New Hampshire, after a man walked into their headquarters with an incredible story. Gerard St. Louis, 51, of Lowell, Massachusetts, said that he and his wife and two sons Alan, 14, and James, 12, were asleep in their truck. The vehicle was parked on the grounds of the local flea market off Route 122, as they were intending to go to the market the following morning. At 10:30 p.m. the family were all asleep when Mr. St. Louis said their pick-up truck began to shake. "I turned around to the boys and told them to stop rocking the truck. They told me they weren't so I decided to find out for myself what the problem was," he said.[143]

When he opened the door, he said the creature was right in front of him — a nine- to ten-foot-tall, hairy form with human features. The "head was twice that of a human head and it had long arms, big eyes, and it came after me," St. Louis said. That's when he leaped back into the truck and drove off. As he did so, he said, "It was looking at us and coming after us. Then we saw it jump a four-foot-high fence and it disappeared."[144] He reported the incident to the police, who searched the area but did not find any indication of the creature.

Mr. St. Louis said he was planning to set up a stand at the flea market the next morning, but was so shaken by the incident that he never returned. Police Chief Paul Bosquet theorized that the family may have been spooked by a bear searching for food from a nearby trash container.[145] St. Louis said, "We never saw anything like it before. It wasn't a bear."[146]

Later that year in July or August, Nancy and John Ingalls were driving home to Clarendon, in east central Vermont, when they spotted a strange creature along Route 7 just south of the junction with Route 103. The encounter happened at 10:15 p.m. as they were returning from an evening of roller skating. They said the creature looked part human, part animal and its eyes glowed in the car headlights. It stood six feet, six inches. Mr. Ingalls drove to the spot the next day and found a trail of over-sized naked footprints fifteen inches long and five inches wide.[147]

During the summer of 1978, a mother and son reported encountering a Bigfoot creature in a wooded region of southwestern Rhode Island near Indian Cedar Swamp. The incident occurred off Route 112 in the town of Charleston in Washington County. At the time they were riding the back roads at 9:00 p.m. hoping to spot deer when a light rain began falling. As they rounded a curve paralleling the Cedar Swamp, they noticed a large oak tree blocking the road from an earlier thunderstorm. They could not get through so they turned around, but this proved challenging due to the narrow road. As an adult, the son described what happened next. As the headlights shone on a tree stump by the roadside twenty feet away, next to the stump they saw "what looked like a large white (yellow white) ape. It was maybe six to seven feet tall, its hair was long, face flat, long massive arms, [and] its head appeared to be without any neck." The chest of the creature was broad.[148]

As for what it was they saw, he says he still finds it difficult to accept. "I'm not sure to this day what I saw wasn't a man in a costume, dressed up, out in a rainstorm, a half mile or more from any home, waiting for my mother and me to drive down the road to give us a scare. The only thing I can say is it was a damn good costume..."[149]

On November 28, 1978, a couple who had recently moved to Shrewsbury in central Vermont to enjoy country life, made an unusual find near their new home on Upper Cold River Road — huge human-like footprints. Mr. and Mrs. David Fretz found the prints after a two-inch snowfall the previous day. Mrs. Fretz was a school teacher in nearby Rutland. When the prints were found the next day, "the dog started growling and smelling the ground..."[150]

In 1978, there were sensational reports out of northern Vermont that a local version of Bigfoot dubbed "Goonyak," had been shot dead in the vicinity of Craftsbury and Morrisville. According to the reports, the farmer shot an eight-foot-tall creature after it dragged a

1,000-pound bull into a nearby field. One story held that it had been shot ten times with a 30.06 rifle. Some versions of the story resembled a B-grade UFO film, and claimed that the creature's death was being covered up by the government, and that armed guards took the body to a secret site on the campus of the University of Vermont in Burlington. Reporter Kevin Duffy of *The Rutland Herald* investigated the claims and found them to be baseless.[151]

The Hockomock Swamp Monster

The swamps around Hockomock in southeastern Massachusetts have been rumored to be haunted by an array of spirits since Indian times, including several sightings of a shaggy ape-man. In 1978, Joe DeAndrade of nearby Bridgewater said he was standing by a pond on the edge of the swamp, known as Clay Banks, when something extraordinary happened. He suddenly had an urge to turn around, and "...[T]here, off to the right, maybe 200 yards away, there was this — well, I don't know what it was. It was a creature that was all brown and hairy, like a big apish-and-man thing. It was making its way for the woods, but I didn't stick around to watch where it was going. I ran for the street."[152] DeAndrade, a former security guard, said the creature "was walking slowly, like Frankenstein, into the brush."[153]

In the summer of 1979, a family were fishing on what the locals refer to as the East Bay River in the remote, isolated town of West Haven, Vermont (which borders the Bigfoot hotspot of Whitehall, New York) when they frightened a huge, hairy creature. The father said his entire family watched as a human-like head peered over some bushes. When the man yelled out, the figure turned and fled eastward. As it moved, he said the head occasionally came into view as it bobbed up and down above the bush line. He said it was hunched over as it ran. They reported a similar encounter a week later near the same spot. This time the man brought a gun, but couldn't bring himself to fire it.[154]

On a chilly winter night in 1983, fur trapper John Baker of West Bridgewater in southeastern Massachusetts, was paddling his canoe quietly along a river in the Hockomock Swamp, setting muskrat traps. Suddenly, there was a loud crashing sound of something running through the nearby woods. Then he saw it. A huge hairy creature entered the river with a splash and passed to within yards of his canoe. He said it had a musty, skunk-like odor. It was his only encounter with the creature despite over three decades of trapping in the area.[155]

No Ordinary Man

The farming community of Tinmouth in east central Vermont was the site of an extraordinary encounter in March 1983 involving a middle-aged couple who were taking a slow ride on a rural road. The man was glancing at a low rocky ridge when he spotted what appeared to be a giant man moving swiftly along it. He said, "The man was very nimble. I couldn't believe how quickly he was moving among the rocks toward the high point on the ridge…My wife and I were suddenly stunned when he stopped and turned facing us. His arms were much longer than a normal man's and he appeared to be much bigger — especially taller — than any man either of us had ever seen." The figure then remained motionless for several minutes, before appearing to look directly at them. Trying to make sense of what he was seeing, he rationalized that it was perhaps a prankster, but quickly ruled it out after what happened next. "Suddenly he raised his giant arms above his head and waved them several times…(then) turned and continued along the ridge with the agility of a gymnast. I was convinced that this was no ordinary man." The creature soon disappeared over the backside of the ridge.[156]

On Monday, August 20, 1983, two Pittsfield, Massachusetts men walked into the offices of *The Berkshire Eagle* to report a strange encounter on October Mountain the previous night. The pair, eighteen-year-old Eric Durant of 183 Field Street, and twenty-two-year-old Frederick Perry of 771 Tyler Street, said they were having a cookout with two friends, when they heard strange noises at 10:00 p.m. By midnight, their curiosity got the best of them and the two decided to see if they could find the source. A hundred yards from the camp, they said they could see the outline of a huge creature fifty yards ahead on the trail. When the cookout finished at 1:00 a.m., their car headlights illuminated the creature in the bushes. It stood six to seven feet tall, had dark brown hair and strange, glowing eyes. Both men emphasized that it stood on two legs and they were certain that it was no bear. As Perry got out of the vehicle and walked toward it, it quickly vanished. Despite its size and encountering the creature in the woods in the dark, he said, "Whatever it was it didn't look like it was going to harm you." The next day Perry and one of his friends went to the site again around noon and said they caught a glimpse of the creature moving very fast and swinging its arms as it moved through the woods.[157]

In 1984, there was a spate of close encounters in east central Vermont. In April, James Guyette was carrying out his morning routine of dropping off bundles of newspapers in the vicinity of Bellows Falls in Windsor County. The time was 5:30 a.m. A light rain was falling as he drove along Route 91 near the Hartland Dam. Suddenly he was startled by a large "animal-man." He said the creature was covered in hair and came "up the bank near the brook, walking fast down the road at an angle, maybe 100 yards away." He said the creature was "tall and lanky with long arms swinging as he walked." Guyette pulled off to the side of the road and tried to make sense of what he had just seen, hoping it would return. The episode made a deep impression on him. Later when describing the incident to his wife, he broke down crying.[158]

One night in the spring of 1984, a longtime Chittenden, Vermont hunter had a frightening experience which he later recounted to fellow resident and Bigfoot investigator Ted Pratt. The man wished to remain anonymous. Pratt relays the account:

> Last spring, a gentleman who is not willing to give his name, woke up during the night with some very loud screaming at his backdoor...he's really not afraid of anything, but he told me...'I just couldn't get out of bed. It was a horrible scream. It lasted...five to seven seconds.' And then he heard his cellar door being ripped-off the hinges. His daughter...[lives] down the road...[and] also heard the scream...We looked over the area and we found one footprint, one handprint...If this is so,

Ted Pratt (left) and Dr. Warren Cook in the Chittenden area.

this has to be the first act of aggression we've ever seen, at least in the state of Vermont.[159]

In late May or early June, an incident was recorded in Hubbardton, in Rutland County Vermont by Bruce Bateau and his mother, Bernard Bateau. While lying in bed at 3:30 a.m., Bruce was awakened by a high-pitched shriek. The pitch was so high he said it was close to a whistle. The following day Bruce found a trail of huge naked footprints in the area where the noises had come from, which was near a thick clump of pine trees. The incident took place just below Monument Hill Road. He also noticed a heavy musty/musky odor — like rotten cologne — which produced a strange feeling.[160]

Haunting Eyes

In November 1984, a family of four spotted a massive man-like creature while driving along Route 2 in Colchester. The incident occurred as the father was driving from Milton to their home in Winooski just north of Burlington, Vermont. At around midnight, a heavy snow squall swept through the area. Suddenly a towering creature crossed the road and began to amble up a hill when it stopped and stared back at the car. The mother said as the creature was looking back at them, it had "one foot on a wire fence holding the wire down with its arm..." Its fur was long and white with dirty yellow streaks, while its eyes appeared amber-yellow, similar to caution lights. The witness said the creature stood at least ten feet tall and had exceptionally long arms. She said its face was covered in what appeared to be fur, with the exception of the cheeks and around the eyes. As the father slowed down to get a better look, their two daughters started screaming and the woman "freaked out," so he drove off. Afterwards she said she "just kept on thinking about its eyes."[161]

There were other Vermont Bigfoot encounters in 1984 that came to light thanks to logging magnate Hugo Meyer of Cabot, one of the state's biggest land owners. Meyer learned of the reports — one in Caledonia County and one in Essex County — through his many industry contacts, though specific details are not known.[162] The year 1985 was a landmark for Bigfoot awareness in Vermont with two scientific conferences held at Castleton State College which garnered national attention. The presentations were led and organized by the college's anthropologist Professor Warren Cook, who developed an interest in the subject while growing up in Washington State. During the 1970s he began to actively investigate sightings in the Green Mountain State, following up on stories he heard through his students and press reports.

The first conference took place on April 24. The response was overwhelming, with an overflow crowd gathering at the Florence Black Science Center to hear firsthand accounts from witnesses and Cook's views on the subject. Cook used the opportunity to call for scientists to conduct research on the creature and to push for a law to protect Bigfoot in the state. The second conference was held on November 25 and yielded similar interest. The publicity from both gatherings, coupled with Cook's stature as a world-class archeologist and popular instructor, led to a flurry of area residents coming forth with their own sightings.

March 4, 1985 was a snowy evening in the Rutland County hamlet of Clarendon Flats. Dorothy Mason and her son Jeff were staying warm inside their house when Dorothy recalled, "I was looking out the window to see how much snow had fallen. I happened to notice some very large and unusual tracks about thirty feet from the window." She called her son and they quickly went out and measured the tracks before they were covered by the snow. They were sixteen inches long, five inches wide and the span between tracks was six feet. Dorothy said the prints looked human and had a "definite big toe and three smaller ones that we could see very clearly and possibly one other." The trail led across the lawn and towards the mountains in a westward direction towards West Rutland.[163]

In June 1985, two men were enjoying a relaxing summer day fishing on Foster Pond in Peacham, Caledonia County Vermont when they spotted a hairy figure on shore. Thinking they were seeing one and possibly two bears, they maneuvered the boat for a closer look. Suddenly, the "bear" stood upright and started walking like a human, before breaking into a run as it dashed into a cedar tree swamp. The startled pair said the bottom of its feet were lighter in color than the rest of the creature.[164]

The events on the evening of September 20, 1985, on a rural farm in West Rutland, would make headlines. The location was the home of Ed and Theresa Davis on old Route 4A, a winding, bumpy road linking Castleton with West Rutland. Bob Davis and Frank "Fron" Grabowski III were visiting the Davises. Fron, an eighth grader at West Rutland High School at the time, was sitting with his friend Bob on the porch at 8:30 p.m. when strange noises were heard nearby. As they walked up a dirt road to see what was making the sounds, they saw a "gorilla-like" creature standing erect, walking towards them. He got the impression that the creature wouldn't harm him so long as it wasn't confronted.[165]

Bob said, "I started walking up the road and saw a black image. It was taller than me and ran like a human." They said it began tossing stones at them, but turned and ran off when the pair began throwing stones back. Bob said the creature was seven feet tall, and he could discern black skin beneath its eyes. As their drama was unfolding, Bob's brother Al circled behind the house to see who was playing a prank on them. He situated himself on a ledge along the gravel road where the creature was moving, with the intention of grabbing the culprit. However, he changed his mind when he saw it, realizing it was no human being. He said it was "humungous in the shoulders...God, to me, it had like a monkey run to it...like a long stride foot...the way it was twisting the shoulders."[166] He said the creature gave off a "gassy or swampy smell" that made him nauseated. He used words like "wicked" and "unreal" to describe it. Several large tracks were found pressed into the hard gravel road where the creature was seen.

This incident may never have been made public if not for two West Rutland teachers, Mel Loomis and Linda Barker, who had been students of Cook at Castleton State College and believed Frank to be trustworthy. After the story circulated through the school, the pair contacted Dr. Cook who quickly went

Cast taken at West Rutland, Vermont, September 1985. There were three casts made of six impressions found after a multiple person sighting.

Paul Bartholomew is seen here with two of the 1985 West Rutland casts at the Castleton State College "Ancient Vermont" exhibit.

to the scene and made plaster casts of the prints. The tracks were fourteen inches long and seven inches wide. Based on the dented gravel, Cook estimated the weight at 400 pounds.[167]

In mid-November nearly two months after the West Rutland encounter, two boys were playing near the Davis home one Wednesday evening at 6:30 p.m. when one of them saw a Bigfoot standing thirty feet away. The witness was holding a BB-gun at the time, and instinctively emptied it in the direction of the figure, then ran for the safety of the house.[168] The following morning, November 21, as the school bus was picking up students it turned around in a nearby parking area. As it did, several students, including Gary and Donna St. Lawrence, spotted a huge, black hairy creature in a field.[169]

A Close Encounter on a Vermont Road

In 1986, there were two excellent sightings in Rutland County. On the evening of October 1, three Castleton State College students — John Bradt, Kerry Bilda and George Dietrich — were traveling along Route 4A from Castleton to West Rutland when their vehicle nearly struck a huge creature on the roadside. Everyone saw the figure which stood six-and-a-half to seven feet tall. Bilda swerved to the left to avoid hitting it. As they looked back, the creature continued to walk west on the east side of the road. By the time they had turned around to get another look, there was no trace of it. Bradt later said that if his window had been down, it was so close to the vehicle that he could have reached out and touched it. He said the body was covered with collie-length hair. The face was virtually hairless, it had deep set eyes, high cheekbones, while the skin tone seemed Caucasian. Perhaps the most curious aspect to the incident was its ambivalence. They said it acted nonchalant and did not try to get out of the way of the car or even flinch as they passed.[170] The incident took place within 300 feet of a Bigfoot sighting the previous year on September 30, 1985, by Susan Cook, daughter of Dr. Warren Cook.[171]

Later that same month, two Castleton College girls spotted a Bigfoot-like creature while driving along Route 30 in the town of Poultney. The incident occurred at 9:00 p.m. on Sunday October 26, halfway between Castleton Corners and Poultney. The girls said they simultaneously screamed when they saw the thing. Jill Cortwright and her passenger Cathy Quill said the creature was hunched over on the side of the road in a crouching position. Jill described the hair as bear-like but was certain it was no bear. The pair were too frightened to turn around and get another look. Quill later told Dr. Cook that the creature's hair was "messy, not smooth like a bear." As the buttocks were facing them, they said the hair was thin in that area. They said the creature seemed to be oblivious to their presence.[172]

In 1987, John Miller was on his farmstead a mile from Rutland, when he reported spotting a ten-foot-tall, black, gorilla-like creature at dusk. Little else is known about the incident.[173]

Also in 1987, a seventy-year-old Webster, New Hampshire hunter stepped into the media spotlight after reporting that he saw a Bigfoot-like creature while pheasant hunting in the nearby town of Salisbury in the south central part of the state. After the incident, he reported his encounter to a family friend — Salisbury Police Chief Jody Heath. Walter Bower said the creature was nine feet tall and its body was covered with grayish hair. While he didn't get a good look at the face, he said, "The hands were like yours or mine, only three times bigger, with pads on the front paws, like a dog." He said the arms and legs were both long, and it looked "like a gorilla, but this here wasn't a gorilla."[174]

When Bower approached the local game warden, it was suggested that he had mistaken a bear or moose for Bigfoot. Bower was adamant that he wasn't mistaken, having shot four bears in his lengthy hunting career. "I said, Look, I can tell the difference between a bear and I can tell the difference between a moose. This was neither."

Concord Monitor reporter Scott French accompanied Bower to the scene in Merrimack County, while writing an article on the incident. He seemed impressed by his character and sincerity, and later wrote, "Now, ask yourself this. Why would Bower, a retired caretaker at the New Hampshire Veterans Home, make up such a story? Why would he subject himself to such ridicule?"

In early autumn 1988, two teenage boys encountered a large creature while walking in the woods of Rutland County. The creature was walking on two legs at a "brisk" rate of speed. They described the creature as standing seven feet tall and weighing 300 pounds. It walked through brush while the boys were on a nearby path. The teen said that his friend saw it first and yelled, "What the ____ is that?" They saw the creature walk by them from a side angle. They estimated it was no more than twenty yards away. After it moved away, the pair raced back to a relative's house nearby.[175]

In December 1989, Guy Primo had taken his wife and two children into the woods on his property near the town of Eden in Lamoille County Vermont when he noticed huge, naked footprints in the snow. He said his hair stood on end after realizing that whatever made the tracks appeared to walk over a small tree that was flattened. He said the tracks were thirteen inches long and six inches wide. The tracks followed a ridge, crossed a pond, and trailed into the forest.[176]

In January 1999, a hunter on a mountain in the vicinity of Ludlow in central Vermont, said he was walking in heavy cover with a friend when he spotted a huge creature thirty yards away. His first reaction was that it was a moose, as he tried to make sense of the encounter. While watching it, he said he suddenly got wind of a "terrible smell, unlike anything I've ever smelled." The creature was partly obscured by a white pine tree so he couldn't see the face. Peering at it through a 5x rifle scope while still at close range, he got a good look at the body. The top of its shoulder was six-and-a-half feet off the ground. He observed its large black head from the back as it "walked away in huge strides." The creature seemed to be covered with a reddish-brown fur that was "matted in places and appeared almost as if shedding." Large "human-looking footprints" were found several minutes later after he checked the area. The hunter was surprised by the creature's agility, climbing "a ridge so steep, it would have been very difficult to follow. Its stride up that slope was easily six feet!" The pair decided that it would be "morally wrong" to try to track and shoot the creature.[177]

Half Ape, Half Human

In September 1999, a man was retiring to bed at his residence in a heavily forested area on the outskirts of Rutland at 2:00 a.m. when he heard "heavy breathing and heavy footsteps on the road outside. I looked out my window and at first I thought I saw a large man in the road. But as my eyes adjusted to the dark I could clearly see a creature about 7–8 feet tall." He estimated the weight between 300 and 375 pounds. "I was staring at it and it locked eyes with me and then I could see portions of its face by the moonlight. Its face had the characteristics of a gorilla but sort of human too." The figure then turned around and walked up a hill, "swinging its arms in a way that wasn't normal," before disappearing.[178]

In May 2000, two cousins were visiting their grandparents who own a wilderness cabin at Marshfield in northeastern Vermont, when they decided to take a drive with their ATVs on an old logging road. The pair were forced to stop when a fallen tree blocked the path. As they were turning their vehicles around, they saw a hairy seven-foot-tall creature rise up and run off into the woods. [179]

Hillsborough County in southern New Hampshire was the site of a reported Bigfoot encounter in May 2002. While delivering newspapers with a car at 4:45 a.m., the witness glanced towards a nearby swamp and saw the silhouette of a Bigfoot-like creature with a cone-shaped head. After quickly getting back into the car and glancing back, the creature was gone. It was said to be hairy, eight and a half feet tall, and black in color.[180]

Encounter on Glastenbury Mountain

In the autumn of 2003, a spectacular sighting of a Bigfoot-like creature was made in southern Vermont. Ray Dufresne, 45, said he was driving along Route 7 in the town of Glastenbury, when he saw a "big black thing" lumbering into the woods near Glastenbury Mountain. He said the creature was at least six feet tall and weighed about 270 pounds. "The first thing I thought was this is a gorilla costume. I thought it was a joke, then I put two and two together." The creature was fifty yards away. He couldn't see its face as it was covered with long, thick black hair. "It was hairy from the top of the head to the bottom of his feet," he stated. "It was not walking like a normal person."[181] The encounter took place near the highest point on Route 7 near dusk at 7:10 p.m.

An ardent hunter since he was 15, Dufresne said it was neither a moose nor a bear. As for the possibility of a prankster in a gorilla suit, why would someone go to such elaborate ends just to scare someone and risk being shot? Newspaper and television publicity surrounding the sighting, prompted other witnesses to come forward with sightings in the area. On September 16, Doug Dorst was southbound on Route 7 at 7:45 p.m. when he was startled by a big six-foot creature with a tan face and a dark body, walking in the road. Dorst assumes that it was someone in a suit out to scare people. On the same evening, two nearby Bennington women, Sadelle Wiltshire and Ann Mrowicki, were traveling north along Route 7 and also spotted the creature. Wiltshire also subscribed to the prank theory: "It was the weirdest thing. It looked like a guy in a gorilla suit, but it looked like it had a tail."

Three hikers reached the crest of Goat Rock near Stoughton, Massachusetts in November 2004, when they detected a bad smell like rotting fish. Suddenly, a large hairy creature ran out from under an overhang, dragging behind it a deer carcass. One of the witnesses said he was "absolutely stunned." At first they started to search for the creature, but after walking fifty yards, they started to get frightened after one of the hikers said he felt an overwhelming sense of danger. The creature had greenish-gray hair, was six to seven feet tall, and had a "broad back." It wasn't in view for more than a few seconds.[182]

Early on the morning of January 1, 2005, a trucker from Duvernay in the Province of Quebec, Canada, was crossing the border into

Orleans County where he and his Vermont girlfriend were going to visit her residence. As they were driving south along Route 243 near Troy, they saw what at first they took to be a bear in the road. The man flicked on his high beams but couldn't see much as the defroster was not working properly and the night was frigid. He stopped his vehicle in the middle of the road to remove a chunk of ice that had accumulated on his driver's side windshield wipers. As he opened the door, he slipped and fell on the ice. At the same time his girlfriend started yelling at him to look at the "bear" that was walking on two legs. His reaction was to start blinking his eyes as the creature kept moving towards his truck, illuminated in the headlights. He scrambled back inside and they locked their doors. He said the creature stood seven feet tall and weighted close to 500 pounds and was visible in the truck headlights for the next several minutes. At first, he said it covered its eyes with one arm. "This arm (not a paw) was black, shagged hair all over and thick but it was hard to make out small details because of the icy windshield." The man, at the prompting of his companion, wanted to drive on, but hesitated as the "thing" remained in the middle of the road. After dimming the headlights, the creature lowered its arm and proceeded to stare at them for several minutes.

The witnesses were worried about the health of the "thing" given the cold, but eventually drove off, fearing for their safety. "You will have to believe me, there was (sic) no clothes on this man thing and on ice the thing was barefooted..." The creature had thick bear-like hair "with some bare spots of hair here and there." At this point, his girlfriend tooted the horn, causing the creature to jump "several feet up straight in the air" before it walked "gimpy-like over to the other side of the road with its back to us." He also noted that the buttocks were "mammoth" and that it walked "funny peculiar."[183]

On the first of September 2005, two hunters were looking for deer in Coos County in Northern New Hampshire. Just before sunrise, they heard a strange noise just ahead of them that sounded like a high-pitched yell. "We looked over," he said, "and walking from behind a bush we saw a dark figure that stood about six or seven feet tall, with a long pointy head." They kept it in sight for ten minutes as it walked down a ridge. Upon investigating the area where it had been, they found footprints between fifteen to eighteen inches long and with a five-foot stride.[184]

On the evening of October 8, 2005, a man was moose spotting with his two daughters near Ludlow in central Vermont, not far from Okemo Mountain, when they were startled by a huge creature walking on two legs. The figure crossed the road fifty feet in front of their vehicle. The man and his daughter sitting in the front passenger seat, both got a good look at the creature. They said it stood about eight feet tall, was covered with hair, and crossed the road in just two strides. The father described the creature as "heavily built" and "covered in short dark hair." He also said that the arms had a pronounced swing as it moved, that its head was cone-shaped, and that the hand that was visible was massive.[185]

In January 2005, a gorilla-like creature "with human features" and standing five and a half feet tall, was spotted walking in the snow near Bristol Connecticut, twenty miles southwest of Hartford. The "thing…had glowing red eyes" and was covered in dark black hair.[186]

On a frosty morning in November 2006 near the tiny, rustic town of Eagle Lake in extreme north central Maine, a hunter who had camped out overnight, decided to walk the rugged, sloping terrain in hopes of shooting a deer. At 7:30 a.m., he heard a grunting sound and, clutching his loaded gun, slowly walked over a hill to investigate. Reaching the top, he lay down, hoping to hear the deer he expected. Peering over the hill, he saw a large, hairy creature that stood nearly eight feet tall. When he made a noise, the creature turned and looked at him for twenty seconds, before running off into the forest. Upon returning to his camp, his belongings were scattered, so he quickly packed up and left.[187]

PART 3 Evaluating the Evidence

6 | What's Going On? The Challenge to Science

> The only means of strengthening one's intellect is to make up one's mind about nothing — to let the mind be a thoroughfare for all thoughts.
>
> —JOHN KEATS [188]

The existence of Bigfoot is an emotional issue, with strong views on both sides of the debate. It is important to approach the subject with an open mind. In our experience of investigating sightings over three decades in New York and New England, we have noted a curious pattern — most people see what they want to see and approach the topic with preconceived ideas. Many believers accept Bigfoot's existence without ever having read a book on the subject, basing their decisions on gut feelings rather than evidence. Conversely, many of the staunchest skeptics dismiss the topic out of hand, without having ever looked at the arguments. This is true of scientists as well as the layperson. The evidence must be examined objectively and unemotionally. This is easier said than done as complete objectivity may be impossible, but it is something to strive for.

Science cannot always offer clear-cut answers; Bigfoot is a classic example. Most scientists are skeptics, but there are a small number of anthropologists, biologists and zoologists who have gone on record in support of the creature's existence. Among them was the late Vermont anthropologist Dr. Warren Cook. He was convinced that Bigfoot roamed New York and New England, and was a pioneer in investigating early sightings in the region. Cook received his share of ridicule from other staff members at Castleton State College until his death in 1989. More recently, Idaho State University anatomy professor Dr. Jeff Meldrum has been chastised for researching Bigfoot. At one point, 30 ISU scientists wrote an open letter critical of Meldrum. However, fellow physicist Trent Stephens defends his work: "Scientists are extremely conservative and have blinders on. Most of the people who have written off his research have never set foot in Jeff's lab, and they are in the same building."[189]

Science considers all possibilities, but ultimately its practitioners–scientists answer questions in probabilities and degrees of confidence. No one can say with certainty that Bigfoot doesn't exist, but clearly most scientists consider it improbable.

Weighing the Evidence

As we sift through the evidence for a community of mysterious apelike creatures on the rural fringes of New York and New England, one is struck by the compelling nature of the reports. There are vivid accounts by police, conservation officers and hunters, several of whom said that they had a clear shot at the creature, but couldn't pull the trigger as it looked too human. Most witnesses seem to be everyday people going about their daily lives: housewives and nurses; lumberjacks and schoolteachers; hikers and cyclists alike. Many remain anonymous, yet have gone to great lengths to make their stories known by seeking out investigators. Witnesses are often haunted by their encounters for decades after as they grapple with the mystery of what they saw. Eventually they too must face the great conundrum: scores of other sightings including multiple witnesses and huge, human-like footprints, yet not one body, fossil or bone to be found. How do we reconcile these contradictions?

Bigfoot theories can be grouped into three broad categories. The physical or "flesh and blood" hypothesis considers the creature to be just another member of the animal kingdom, albeit elusive. Social explanations view Bigfoot as belonging to the world of folklore — a world that is driven by media sensationalism, rumors, misidentifications, half-truths and wishful thinking; a creation of the human mind. Lastly is the paranormal or "Twilight Zone" hypothesis. Could Bigfoot originate from another dimension or as a psychic projection from our subconscious? Each of these explanations rest on certain assumptions as to the nature of the world and how it works.

Flesh and Blood Explanations

Among believers, this is the most popular explanation. Over the past two centuries, zoologists have been startled by the discovery of a plethora of new species that were either once thought to have been extinct or had never been classified. The discovery of the first gorilla by Europeans, has obvious relevance to the Bigfoot mystery. During the mid-1800s, there were several unconfirmed sightings of these hulking creatures, but the accounts were widely dismissed as the products of overactive imaginations. Doubt gave way to acceptance in 1856 when explorer Paul du Chaillu shot one dead. The story does not end there. During the latter decades of the nineteenth century, there were rumors of a super-sized gorilla living in the northern Congo. Once again the accounts were widely ridiculed as the embellishments of excitable people or outright fabrications. In 1902, fiction gave way to fact when German military officer Oscar von Beringe was climbing rugged Mt. Sabinio and became the first European to see the mountain gorilla *(Gorilla beringei beringei).*

The okapi of the Congo rainforests *(Okapia johnstoni)* was not detected by Europeans until 1901 when British naturalist Sir Harry Johnston spotted one in the Belgian Congo. This deer-like relative of the giraffe, resembles a cross between a zebra and a horse. It took nearly twenty more years to finally capture a specimen. Shy and nimble, it dashes into the thickets at the first inkling of danger. The existence of the Ethiopian mountain nyala *(Tragelaphus buxtoni)* which can weigh over 600 pounds, wasn't established until 1910. The pygmy hippopotamus, Komodo dragon, congo peacock, and Coelacanth round out some of the major discoveries through 1940.

The congo peacock *(Afropavo congensis)* wasn't discovered until 1936 in the Sankuru region of the Congo; several expeditions since have failed to find any trace of the elusive bird, but we know they exist as some have been captured and bred in captivity. The pygmy hippo *(Choeropsis liberiensis)* is found only in a small swampy area of northwest Africa. Despite weighing up to 600 pounds, they were not identified by Europeans until the early twentieth century.

Perhaps the most spectacular finds during this period were the Komodo dragon and the Coelacanth. During the early twentieth century, there were reports of giant dragons on the tiny volcanic Indonesian island of Komodo near the equator. Then in 1912, scientists were both shocked and elated when Europeans found the world's largest lizard, soon dubbed the Komodo dragon (*Varanus komodoensis*). Spanning nearly ten feet from head to tail, it is known to kill deer, pigs and even people. The reptile was subsequently found on several nearby islands. Another spectacular discovery was the Coelacanth *(Latimeria chalumnae),* a strange-looking fish thought to have become extinct sixty-five million years ago. When fishermen caught a specimen in their nets off southern Africa near the Comoro Islands in the Indian Ocean in 1938, it was dubbed "the living fossil." Prior to its discovery, the earliest record of existence were fossils that were eighty million years old. This has direct relevance to Bigfoot as a major objection by skeptics is that no fossils have ever been found.

The gigantic megamouth shark *(Megachasma pelagios)* was only discovered in 1976, yet can weigh over a ton. In 1990, the black-faced lion tamarin *(Leontopithecus caissara)* was found on the populous Brazilian island of Superagui just 40 miles from the mega city of Sao Paulo. The creatures survived and eluded detection even though 97 percent of their forest habitat had been lost to human encroachment. Russell Mettermeier, head of Conservation International, was astounded by the discovery, noting that "it's almost like finding something in the suburbs of Los Angeles."[190] Four years later Australian zoologist Tim Flannery established the reality of the Bondegezou, a new species of tree kangaroo *(Dendrolagus mbaiso)* in the remote rainforests of Papua New Guinea. That same year, an Australian park ranger vacationing in a remote corner of Wollemi National Park, stumbled upon thirty-nine pine trees covered in a strange, waxy brown bark resembling chocolate bubbles. Named the Wollemi Pine, or *Wollemia nobilis,* the trees were thought to have become extinct 200 million years ago. Sydney botanist Carrick Chambers said, "The discovery is the equivalent to finding a small dinosaur still alive on earth."[191]

In 2003, scientists announced the discovery of a new species of African monkey. The acceptance of the highland mangabey *(Lophocebus kipunji)* was finally confirmed in southern Tanzania after a series of sightings, photos and recordings of distinctive vocalizations.[192] In late 2005, a scientific expedition to the isolated Foja Mountains of northwest Indonesian New Guinea, uncovered a "lost world" with seventy new species — some thought to have been extinct — twenty new frogs, four new butterflies, and several new plants including a flower that may be the world's largest rhododendron.[193]

If all of these creatures have eluded detection for so long, why couldn't Bigfoot? Many were mammals of recent discovery and large in size. But wouldn't such creatures be discovered by now? Perhaps not. In January 2007, Sheriff's Deputies in Warren County, New York arrested a lean, muscular man who told an incredible story of surviving in the remote woods of Upstate New York, New Hampshire and Maine for nearly twenty years as a modern-day mountain man living in make-shift shelters. During this period he was only caught twice — for burglarizing camps. He mostly took items he needed to survive, like food and camping gear, and he never hurt anyone. He eluded police in Grafton County, New Hampshire for several months in 2002. Police detective Ken May said, "It was an amazing case...We chased him and chased him. The guy has phenomenal survival skills."[194] When he was caught in 2007, Sheriff Larry Cleveland was incredulous. "We used one of our global (background) searches on him and found almost nothing — no prior addresses, no relatives, no taxes, licenses, nothing." He said his last driver's license was issued in 1986, and since then he "seems to have fallen off the face of the Earth."[195] If a mere man could do this, could Bigfoot manage to elude capture if it had evolved specific mechanisms to keep it hidden away?

If Bigfoot does indeed exist, what could it be? Theories are varied but researchers point to two most likely candidates, both large ape-like creatures that are now extinct.

The Prime Candidates

The late Washington State University anthropologist Dr. Grover Krantz was convinced that Bigfoot is a relative of the giant ape *Gigantopithecus blacki* who roamed what is now China, India and Vietnam as recently as 200,000 years ago. *Gigantopithecus* is Greek for gigantic ape, of the species blacki, after anthropologist Davidson Black. Sometimes nicknamed Giganto, it resembled King Kong as it towered above

Dr. Grover Krantz

the landscape at over ten feet tall, and weighed upwards of 1,200 pounds. Canadian wildlife biologist Dr. John Bindernagel holds a similar view. "It's obviously...a big, hair-covered mammal...There are other mammals on Earth and there is this group — the great apes. And those are the closest ones. Now granted, they don't occur in North America. They're basically tropical in Africa and Asia. But at least we can look at what seems to be a relative or related animal elsewhere on the planet as a possible explanation for what we're dealing with here."[196] Dr. Bindernagel believes that the sightings in the eastern states are just as convincing as those in the more well-known Pacific Northwest. After pondering how an eastern Bigfoot could find enough food to survive, he came to the conclusion that it was very possible. "I realized that the deciduous (seasonal trees) forests of the East are in many ways richer in life than the coniferous (evergreen) forests in the West, and that could sustain a large ape."[197]

In 1985, Liu Minzhuang, Professor of biology at Shanghai's China Normal University, concluded that Chinese Bigfoot sightings (referred to as the yeren, or wild man), were probably related to a species of *Australopithecus* (southern apes) that once roamed southern Africa as recently as a million years ago. Dr. Warren Cook also believed that witness descriptions around the globe most closely resemble *Australopithecus*, rejecting the Giganto theory as most primatologists agree Giganto walked on all fours. Cook believed that witnesses are seeing "a cold-weather-adapted swamp adapted Australopithecine,"[198] which, he thought, explains their presence in the northeast, as much of the region is either forest or bush.

Copy of a yeren footprint cast obtained by Dr. Warren L. Cook when he visited China in 1985 to exchange information on Chinese and North American Bigfoot-like creatures.

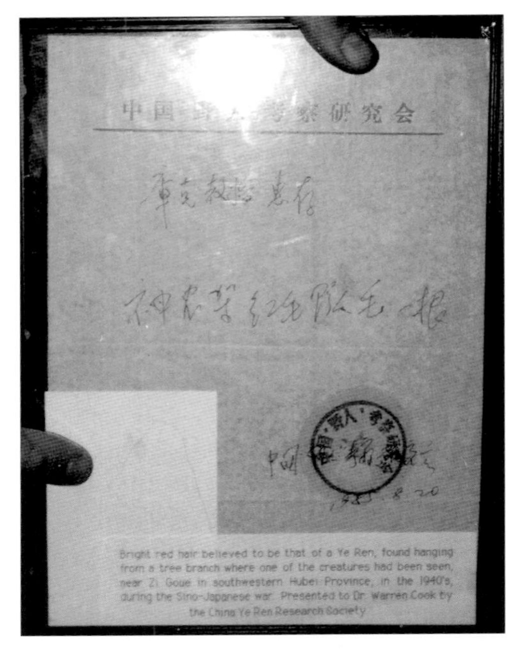

This unusual artifact was also presented to Dr. Cook when he visited China. The information shown reads: "Bright red hair believed to be that of a Ye Ren [yeren], found hanging from a tree branch where one of the creatures had been seen, near Zi Goue in south western Hubei Province, in the 1940s, during the Sino-Japanese war. Presented to Dr. Warren Cook by the China Ye Ren Research Society."

Anthropologist Colin Groves of Australian National University says a major problem with the existence of Bigfoot is fingering a plausible candidate. Groves says the line-up of possible suspects is limited and not particularly convincing as none have a recent fossil record. "While absence of evidence is not evidence of absence, it must be taken into account that, if Bigfoot exists, there is so far simply nothing for it to be descended from."[199] There is no fossil record of non-human primates living anywhere in North America. This would not only make Bigfoot unique in the animal kingdom, but seemingly defies logic. How could a creature exist and be sighted in so many places without anyone finding even a single fossil? True, the earliest fossil record of the Coelacanth was eighty million years ago, but when it comes to Bigfoot, we should expect a recent hominid fossil record of a creature resembling modern-day Bigfoot reports. There is none — at least not yet.

While Groves admits that some witness descriptions resemble reconstructions of Giganto or *Australopithecus*, these creatures died out hundreds of thousands of years ago. Furthermore, any reconstruction of Giganto is highly speculative as everything known about the creature is based on findings of teeth and three mandibles.[200] But again, what about the Coelacanth? On the other hand, this fish is known only in a few places in the ocean, whereas Bigfoot is seen on land and there are Bigfoot hotspots all over.

Psychologist Eric Pettifor of Simon Frasier University is also skeptical that a descendent of Giganto is what people today are reporting as Bigfoot. He observes that a creature of such size, and existing in such numbers as to maintain an ongoing breeding population, would have a noticeable impact on their environment.[201] Such a large creature would require a massive amount of food.

Footprint Evidence

Some individual prints are tantalizing and have been deemed authentic by a handful of scientists. Dr. Grover Krantz studied Bigfoot footprints. He states that on authentic prints, a small "push off" mound of soil is created when the creature pushes horizontally off its forefoot just before it lifts off the ground. He says he has tried many times to duplicate this feature using either wooden or rubber feet, but was unsuccessful.[202]

Jeff Meldrum, associate professor of anatomy at Idaho State University also believes the footprint evidence is convincing, based on his collection of over 200 plaster casts of supposed Bigfeet. While admitting that some are obvious hoaxes, he believes that most are consistent with primate physiology: feet that are flat, flexible and have five toes.[203] Dr. Bindernagel also believes that based on footprint evidence alone, Bigfoot must exist. As for the huge variation in print sizes, he is not dissuaded. "If you were in a locker room and saw all the bare feet, there'd be a lot of variation," he says. [204]

James Chilcutt, a professional latent fingerprint examiner, former FBI agent, and real life crime scene investigator (now retired), says the more he looked into the alleged Bigfoot prints, the more convinced he become that they were those of a natural animal. Now there is no longer any doubt in his mind. "The evidence as I've examined is proof-positive to me that there is an animal out there — some sort of primate running around..." What is so compelling about some of the casts he has analyzed? "I started to examine them and noticed the actual friction ridges that ran lengthwise down the foot, and the dermal ridges texture was about twice the thickness of the human. In humans, the ridge pattern goes across the width of the foot. In this cast, they go down the side of the foot." Chilcutt says he sees the normal ridge pattern in his everyday work. That's why many of the so-called Bigfoot prints stand out and were not likely hoaxed. "When we see a ridge pattern flow that's completely opposite of human beings, yes, that is kind of shocking."[205]

Many mainstream scientists who have looked at the footprint evidence, find it fascinating, but say that it still falls short of proof, and may represent wishful thinking. They have come to the opposite conclusion as Krantz, Bindernagel, Meldrum and others — that no clear pattern emerges. There are prints with two toes, three, four, five, six and even seven toes! Some have heels while others don't. Some have thumb-like big toes that resemble apes; others don't. Some toes are dramatically splayed and resemble giant duck feet; others are close together. Then there is the question of distinguishing the genuine article from the fake. Dr. Krantz was once fooled by a series of footprints leading up a steep, snowy hillside and concluded that it could not have been made by a person. He was wrong. "I had to admit that no person could have run up that slope with eight-foot steps, fake feet or not. It was later found that a high school athlete had made the tracks; he wore fake feet that were put on backwards, and he ran down the slope."[206]

Hair and Film Evidence

Bigfoot hair has been found all over the world, but to date nothing is conclusive. However, there have been some interesting cases. Researchers recently found a hair at the base of a cedar tree where a yeti, or abominable snowman, had been sighted. The hair was given to Professor Bryan Sykes of the Institute of Molecular Medicine at Oxford University, who was surprised by the results. "It's certainly mysterious…there was a plump follicle there. We normally wouldn't have any difficulty at all. It had all the hallmarks of good material…We found some DNA in it, but we don't know what it is." Of course, this doesn't prove that the hair came from a Bigfoot, but his findings do raise eyebrows. "It's not a human. It's not a bear. It's not anything else that we've so far been able to identify — a mystery. We've never encountered any data that we couldn't recognize before..."[207]

This isn't the only time such findings have been reported by scientists. Hair collected from the scene of a Bigfoot close encounter in remote Pike's Peak, Colorado, was sent to a top analyst in the field of hair identification, Dr. Jerold Lowenstein at the University of California at San Francisco. He too was surprised by the results. "I've tested these hairs for all the major groups of mammals that have large specimens — like deer, bear and so forth, and it only reacts to the primates. And of the primates, it only reacts with hominids. And there are only five hominids: human, chimpanzee, gorilla, orangutan and gibbon." Lowenstein believes that the hair belongs to a "large animal closely related to human and chimpanzee," but prefaces: "it's hard to see how this could have escaped detection over thousands of years."[208]

A key problem with identifying Bigfoot hair is that you need a specimen to match it to. The only thing you can conclusively prove is that it wasn't from Bigfoot. Many books and documentaries claim that suspected hair was analyzed and could not be identified as belonging to any known animal. This does not necessarily suggest that it was from an unknown animal.[209] Even Dr. Krantz, a Bigfoot believer, was cautious on the subject of hair. "When a hair cannot be matched, labeling it as an unknown species is not necessarily warranted" because there is no comparative collection of every type of mammal hair. "A hair that is unlike anything in a North American collection might be from the armpit of a bear or from an escaped llama," he said.[210] The science of analyzing hair can be open to much interpretation and is far from exact. Krantz states that numerous hairs purported to have been from Bigfoot have been found throughout the American Blue Mountains and were sent to various experts for analysis. "Most of these called it human, the Redkin Company found significant differences from human hair, but the Japan Hair Medical Science Lab declared it a synthetic fiber." Further complicating matters, a scientist at Washington State University concluded that it was an unknown type of hair. Eventually the sample was positively identified by European scientists as Dynel, a substance commonly used to make artificial hair.[211] Clearly, hair analysis is an inexact science.

Footprint finds, hair samples, films and photos (which tend to be suspiciously blurry or crystal clear) will not settle the argument. Nothing short of a body (or body part) will do.[212] Scientists are only human, and as such, are prone to interpreting evidence to reflect their personal beliefs. For instance, on October 20, 1967, the most famous Bigfoot film ever shot was taken by Roger Patterson and Bob Gimlin near Bluff Creek, California. The pair claimed to have stumbled across a Sasquatch in the woods. Dr. Grover Krantz was convinced the footage was genuine and not someone in an ape costume, remarking that the creature's gait could not be copied by a human. "The size and shape cannot be duplicated by a man, its weight and movements correspond with each other and equally rule out a human subject; its anatomical details are just too good."[213] Dmitri Donskoy of the Central Moscow Institute of Physical Culture, supported Krantz, concluding that the image in the film was "a very massive animal that is definitely not a human being."[214] Anthropologist Dr. John Napier, a former primate biologist with the Smithsonian Institution, saw the same film and came to the opposite conclusion — that it was a man in a costume. He said, "the presence of buttocks, a human hallmark…is at total variance with the ape-like nature of the superstructure…The upper half of the body bears some resemblance to an ape and the lower half is typically human. It is almost impossible to conceive that such structural hybrids could exist in nature."[215a] Belgian zoologist Bernard Heuvelmans and University of Florida anthropologist David Daegling also believe it was a man in a suit.

Brown University Biologist Christine Janis counters claims that the creature in the Patterson film could not be an ape because it has human-like buttocks. "It's not a logical argument to claim that because Bigfoot looks like no known ape that it can't be any type of unknown ape. Should another ape (besides ourselves) have evolved the habitual bipedal walking, then they would also have likely evolved methods of stabilizing the mass of the trunk over the hips, such as enlarging the gluteus maximus muscles as in humans, which is what gives us our larger buttocks. Note that members of the horse family, which habitually fight by standing on the hind legs, have larger gluteal muscles than other hoofed mammals, and correspondingly larger, more rounded buttocks. A Bigfoot with a big rear end is actually a predictable likelihood rather than an impossibility!"[215b]

Paranormal Explanations

Some researchers propose that Bigfoot comes from another dimension, which from time to time interfaces with our own universe, offering fleeting glimpses of a different realm of existence. Others suggest that it is a psychic projection from our subconscious that temporarily takes on a reality of its own. These theories can explain the sightings and physical evidence such as hair, footprints that suddenly stop in the middle of nowhere, and the absence of a body, fossils or bones. This explanation could shed light on a relatively small number of bizarre reports that have often been dismissed by investigators who assume its flesh and blood nature — reports where the creature was described as having been transparent or disappeared in a flash of light. Were these witnesses hallucinating? Drinking? Mentally disturbed? Were they observing a creature from another dimension? Of course, the problem with the interdimensional and psychic projection theories is that they break the known laws of physics. To say that these and other kindred theories are highly speculative in the eyes of most scientists, is an understatement. To suggest that eggplant sales will rise by twenty-five percent next year would be speculative but plausible. But to suggest that eggplant will increase due to a sudden, unexpected demand from space aliens is highly improbable. Of course, it is theoretically possible that space aliens exist; they may even make contact with Earthlings this year, and they may have a hankering for eggplant. However, the odds against this series of events happening are astronomical. And, for that matter, what is another dimension? On the other hand, the laws of physics as scientists understand them, may change radically in the years to come.

The Geomagnetic Hypothesis

Parapsychologist Dr. Alan Vaughan and paranormal researcher Peter Guttilla have proposed a fascinating theory about the origin of Bigfoot, postulating that they could be linked to the earth's geomagnetic field. Vaughan and Guttilla claim that sightings of strange creatures, UFOs and reports of ESP, increase when the geomagnetic field decreases in activity. Their theory goes something like this: as the earth rotates, a geomagnetic field is produced as the molten liquid spins in the outer core. This field continuously fluctuates and is sensitive to the powerful peaks and troughs of the sun's own geomagnetic field and solar storms. When the field is low, a parallel universe begins to interface with our own. Vaughan and Guttilla believe that the earth's geomagnetic field could act as a natural barrier that protects us from too much interaction from a parallel universe in another dimension. They set out to test this hypothesis by examining changes in geomagnetic activity which are measured at three-hour intervals from monitoring stations around the globe, and averaged together as a single reading by the National Oceanic and Atmospheric Administration in Boulder, Colorado.

Of fifty-seven high quality Bigfoot sightings analyzed, 77 percent occurred "during times of negative geomagnetic shifts from the day before." They believe that this can explain the great conundrum that we have mentioned throughout this book. As Vaughan and Guttilla observe, "They have managed to outrun, outdistance and generally outwit their would-be captors with unearthly regularity. Dozens of recorded cases tell of footprints leading into meadows and open fields and abruptly stopping as if the creature simply vanished. Others describe the creatures' sudden disappearance in a flash of light or nearby humming and eerie sounds, or the presence of hovering UFOs."[216] They correctly note that most Bigfoot researchers have ignored these high strangeness cases over the years as they assume that Bigfoot is a flesh and blood creature. It is common to assume that witnesses to such bizarre events were mentally disturbed or making it up, but they suggest another explanation. "These thorns in the side of conventional research constitute the single most significant clue to the true origin of Bigfoot. It is our contention that Bigfoot is a visitor from elsewhere."[217]

Psychic Projections?

What about the psychic projection theory? If Bigfoot is a temporary projection of the human mind based on some cultural expectation, this could explain why people once reported seeing fairies by the thousands, but today they see beings that are more plausible such as Bigfoot and extraterrestrials. But if this explanation is true, why aren't there scores of sightings by children and adults of Santa Claus, The Easter Bunny and Tooth Fairy? If young adults believe in the Easter Bunny and are capable of creating a temporary psychic projection, why aren't there more snow-covered lawns filled with giant rabbit tracks on Easter morning, and corroborative sightings by adults?

Of course, no one knows what new discoveries await. During the twentieth century, humanity progressed from the Model-T to the space shuttle. Science considers all possibilities, but in the end, data is interpreted based on probability. If there is a paranormal explanation for the strange world of Bigfoot, understanding it at this juncture in our scientific infancy may be so beyond our feeble brains that it would be the equivalent of a grasshopper trying to comprehend the Internet.

Movie producer Bruce Hallenbeck, whose relatives have reported numerous Bigfoot encounters in rural Kinderhook, New York, was once a proponent of paranormal explanations but now thinks differently. During periods of local Bigfoot activity, Hallenbeck has found three-toed tracks in the middle of nowhere that suddenly stop, as if a creature had dematerialized. On another occasion he heard eerie monkey-like chattering, then saw a strange light in the sky. "Perhaps the three-toed tracks had nothing to do with Bigfoot; perhaps they belonged to some bird that took off into the air, and perhaps the monkey-like chatter was made by something else. Perhaps the light in the sky was merely a coincidence." By the same token, in recent times many weather events have been blamed on global warming, whether they are related or not. Hallenbeck now subscribes to the flesh and blood theory, noting that "the eyewitness sightings have been too consistent, the worldwide belief in such beings, too ubiquitous. There has to be some foundation in our reality, not in some quasi-mystical quantum universe."[218]

Social Theories

All social explanations of Bigfoot consider the origin to be the human mind. Obviously, some cases are hoaxes, but probably no more than 5 or 10 percent. Why hoax? Financial gain, notoriety and a feeling of self-worth are probably the top reasons. Most witnesses seem to be honest, sincere people who genuinely shun publicity and wish to remain anonymous. Most Bigfoot researchers we know are well meaning and equally sincere. However, like the study of UFOs and ghosts, the field of Bigfoot research is not without its hoaxers and psychologically fragile. We distinctly recall talking to one man who told us he had decided to devote his life to proving the creature's existence when, during a bus trip, he looked up in the sky and saw a cloud in the shape of a giant foot. He believed that god was giving him a sign, and that was why none of the other passengers could see it.

Misperceptions of real animals or objects viewed at a distance, underpin social explanations. The Bigfoot phenomena may be a common human experience that represents a desire to believe in monsters. Ask even die-hard skeptics, and most will tell you that they hope Bigfoot exists. The human need to believe, combined with the notoriously poor nature of human perception, may account for many sightings, especially those from afar. Human perception is highly unreliable[219] and influenced by one's mental set at the time.[220] Honest people make mistakes all the time. Professional umpires and referees have excellent eyesight, are trained observers and constantly hone their skills. Yet, it is nearly impossible to watch a pro football or baseball game without instant replay showing one and often several questionable calls. In the end, investigators are left with no convincing physical evidence that could be used to prove the creature's existence: fossils, bones, a carcass, DNA. Even tantalizing hair evidence is inconclusive.

While it may be tempting to conclude that many witnesses were crazy, had too much to drink or were smoking marijuana, many Bigfoot sightings are explainable using mainstream theories of social psychology. Just how fallible is the nature of human perception? A page from the history of UFOs can help to illuminate the process. The mind does not work like a video recorder that simply takes in information. The mind interprets information based on preconceived beliefs and the context, such as being alone in a dark wooded area and hearing a strange, frightening shriek. At 8:45 p.m. Central Standard Time on March 3, 1968, the Russian Zond 4 moon probe plunged back into the Earth's atmosphere resulting in several man-made meteors across the continental United States.[221] After witnessing what was undoubtedly the re-entry, one witness told Air Force investigators that there were "square-shaped windows" and that the "craft" had a "riveted-together look." Further, "many windows seemed to be lit up from the inside of the fuselage...I toyed with the idea that it even slowed down somewhat, for how else could we observe so much detail in a mere flash across the sky?" The conclusion of the witness and two other observers in the group: they had seen a spacecraft from another planet.[222]

Since someone's mental outlook at the time of a sighting is influential, the context of the episode is of great significance. Almost everyone has heard of Bigfoot. Hence, when many people go hunting or for a hike the woods, it could be argued that they are in a Bigfoot state of mind. Of course, most people who walk through the woods aren't expecting to see Bigfoot, but under the right circumstances, Bigfoot may be considered an explanation. By the same token, when we are alone in a creaky house late at night and hear a strange noise, many people would become more frightened than if the noise had occurred in broad daylight. People become frightened and start thinking of and looking for ghosts. If someone is alone in the woods at night and sees a meteor fall, they might automatically think of UFOs. As one team of psychologists has noted, in ambiguous settings (such as people scanning a forest for the existence of Bigfoot) "inference can perform the work of perception by filling in missing information in instances where perception is either inefficient or inadequate."[223]

Adirondack historian Fred Stiles describes a Bigfoot encounter involving neighbor Hail Hall, who was prospecting for lost coins and jewelry in the woods at a time when there had been a spate of local sightings. Hall was bent over digging when he looked up and "saw a giant form about 10 feet tall with a great bushy head. He threw the metal detector down and went galloping off at great speed till he had gone some distance." Hall then realized that he had left his expensive machine behind. Overcoming his fear, he slowly crept back to the spot where the encounter had taken place. "Glancing towards the place where Bigfoot had been, there he was and he was wiggling his head back and forth — it was a very large porcupine, munching on a limb about 10 feet up a tree. And I guess he believes as I do, if the people who see Bigfoot had time to look more closely, they would find something which could be explained."[224] Fear and excitement have long been known to cause the imagination to run wild and lead to misperceptions. As Shakespeare wrote, "Or in the night, imagining some fear, How easy is a bush suppos'd a bear!"[225]

In 1986, a veteran hiker trekking in the Himalayan Mountains, claimed to have spotted a yeti standing by a ridge that was no more than 500 feet away. The creature appeared to be at one with its surroundings as it stood still and didn't make a sound. He said the yeti was "covered with dark hair" and its head was "large and squarish." He also noticed a set of possible footprints leading off in the direction of the figure. Tony Woodridge snapped two photos that were later examined by analysts who concluded that the pictures had not been faked. Some anthropologists, including Dr. John Napier, were excited about the possible encounter. Many Bigfoot researchers suggested that Woodridge had spotted the legendary yeti, pointing to his hiking experience and familiarity with the area. The incident was so compelling that the following year researchers traveled to the spot to investigate further. They determined that Woodridge's yeti was actually a rock. Upon learning of the findings, Woodridge readily admitted to having been mistaken.[226]

Could Bigfoot be entirely folklore? Throughout history, in every culture, people have told tales of seeing mythical creatures such as fairies. Over the centuries there have been thousands of eyewitness sightings of fairies, often in great detail. In many instances there were multiple witness encounters. Assuming that fairies do not exist, is it possible that the same myth-making process that was responsible for the rise of fairylore in recent centuries, has given rise to a Bigfoot myth? If true, Bigfoot could be considered an overgrown fairy that appears plausible to a more modern audience.

High on the list of influential processes in the potential creation of Bigfoot lore, is the media: books, magazines, newspapers, internet blogs, TV documentaries, films, and syndicated radio. There are dozens of Bigfoot organizations devoted to proving its existence. Conversely, there was once a massive literature on fairies, numerous magazine and newspaper reports on fairy sightings, and even groups trying to prove the reality of The Little People. In England, there was once an organization devoted to documenting fairy sightings called The Fairy Investigation Society headed by Leslie Shepard. How could so many people once believe in fairies and claim to see them by the tens of thousands? No one ever found a fairy body, fossils or bones — just as there remains no conclusive proof of Bigfoot. The evidence is either in the form of vague photos and film images, footprints and/or witness testimony. Coincidentally, there were numerous alleged pictures of fairies over the past two centuries, numerous eyewitness accounts and even many reports of tiny footprints in the snow and mud — all said to be proof that fairies were real. Even the creator of Sherlock Holmes, Sir Arthur Conan Doyle, was fooled by the infamous Cottingly fairy photos taken by two young girls in Cottingly, England in 1917. The pictures created a worldwide sensation and for decades after, were cited as evidence for the existence of fairies. In 1981, one of the girls admitted to hoaxing the pictures using a box camera, officially unmasking the deception. Given that Bigfoot has never been found, it is plausible to consider that it may be an example of living folklore, driven by media sensationalism, misperceptions, hallucinations and tall tales.

Another piece of evidence weighing against the flesh and blood theory, is "the numbers game." Simply put, there are too many reports in too many places. There are hundreds of Bigfoot hotspots all over the world and sightings from every state in U.S. It defies logic and the laws of probability that it could elude capture in so many places for so long. Today there are over seven billion people on the planet, yet not one has shot Bigfoot while hunting and brought back the body. Not one person has stumbled across a body in the woods, when it did not mysteriously disappear. Not one fossil or bone. If Bigfoot is a flesh and blood organism, this state of affairs seems to defy logic. Yes, there have been many remarkable discoveries in recent decades, but it can be argued that Bigfoot is different from finding a giant squid in the Atlantic, a new species of bird in New Guinea or a large mammal such as the okapi in the Congo. Bigfoot is reported all over the world and not in some isolated, sparsely populated corner of the earth. If it does exist, there cannot be just one or two hiding on some remote mountainside. There must be small breeding populations. As V. Rae Wigen observes, "Since no tools have been found they presumably would have to survive by gathering or scavenging; this might be possible in a coastal ecosystem but would be extremely difficult in the boreal forest, especially in winter."[227]

On the other hand, Australian researchers Paul Cropper and Tony Healy suggest that Bigfoot may emit ultra-low sounds that are inaudible to humans, which disorientate or frighten potentially threatening creatures. They point to research on attacking tigers which emit frequencies below 20 cycles per second — inaudible to humans but capable of inducing feelings of fear and confusion. They suggest that a similar mechanism could explain why witness-

Australian yowie researcher, Paul Cropper (left) and Dr. Robert Bartholomew in Australia, 1988.

es often feel they were being watched prior to their sighting, and why birds and insects are commonly noticed to go quiet, while potentially threatening animals such as dogs, often go berserk and cower in fright? Many witnesses in New York and New England report another curious pattern — the creature seems to take no notice of them as if they were not even there. Is it possible that Bigfoot has developed some sophisticated defense mechanism whereby it need not be concerned about the presence of humans and other mammals? [228]

The Final Word

If Bigfoot is out there, someone, somewhere will eventually find hard evidence: a body, fossils, bones. At present, the overwhelming majority of evidence is soft: photographs, sound and video recordings, footprints and eyewitness testimony. No matter how compelling, each will always fall short of proof that the scientific community requires. To confirm the existence of a Bigfoot hair, one has to have a body to match it to. Until the day Bigfoot is found, scientists have no choice but to be cautious. The challenge for investigators is not to try to change scientists' minds by citing anecdotal evidence such as eyewitness accounts, for as much as they want to proclaim Bigfoot's existence, they are bound by the oath of science. If Bigfoot is a flesh and blood creature, we can be confident that scientists will soon confirm its reality. With billions of people walking the planet, and civilization encroaching on wilderness areas, it is inevitable that Bigfoot will be found. With each year that passes without the creature being caught, scientists cannot be blamed for growing more wary.

We must set aside our personal feelings and allow science to resolve the debate. The wheels of science turn slowly, but they turn. The history of meteorites is a classic example. For thousands of years people reported seeing stones fall from the sky. Some even claimed to have physical proof — the stones themselves. Many early explanations had religious overtones. Like the Bigfoot enigma, it was a global phenomenon, and witnesses were sometimes ridiculed. Scientists knew that stones did not fall from the sky. Most reports were dismissed as tall tales by the ignorant and impressionable. But with the accumulation of more and more credible sightings — in some cases hundreds of eyewitnesses — an explanation was in order. Among the possible causes: rocks expelled by volcanoes, rocks sent hurling through the air by hurricanes, and rocks created through some mysterious process involving lightning bolts. Then in 1794, German physicist Ernst Chladni published a short book in which he postulated that flying stones originate from space and heat up by friction. Chladni came to this conclusion after studying eyewitness accounts from around the world. Still scientists scoffed. It wasn't until 1803 that a majority in the scientific community were forced to accept Chladni's theory after the respected French scientist Jean-Baptiste Biot issued a report on the spectacular disintegration of a fireball which resulted in thousands of stones raining down near the French village of L'Aigle. There are many parallels with Bigfoot — credible witnesses giving relatively uniform descriptions of something that was yet to be accepted by science. In the end, the existence of meteorites from space became accepted into mainstream science. Will Bigfoot follow suit? [229]

If Bigfoot turns out to be a creature of folklore originating in the human mind, we should embrace this tale for the valuable lessons it offers about ourselves and the human propensity for mythmaking; why we need our monsters. The persistence of reports can also be viewed as an anti-scientific symbol in an age when science has eroded our religious traditions and beliefs. It may be a subconscious urge to rekindle spirituality and preserve our environment when the products of science and the Industrial Revolution have placed the very existence of planet Earth in jeopardy.

There is also a sense of excitement and adventure in joining the hunt for Bigfoot, opposing professional biologists and anthropologists, most of whom tell us it doesn't exist. As Peter Dendle observes, "To be on to something that even the professors of Harvard do not know about, or to benefit from a cure of which the National Institutes of Health are ignorant, can be very empowering in an age of routine deference to higher bodies of institutional knowledge."[230] To be a Bigfoot hunter, all one has to do is trek into the woods or scan the roadsides, or depending on where you live, your own backyard, and look around. To do so is to become involved in one of the greatest potential discoveries in human history. Dendle believes that the search for Bigfoot and similar "hidden animals" may represent "a quest for magic and wonder in a world many perceive as having lost its mystique."[231] Rochester New York cultural anthropologist Ernestine McHugh concurs. "A lot of people are disenchanted with everyday life, and they long for something that seems magical. We've even domesticated Halloween as a day for children, so for many people, Bigfoot is one of the few avenues left to the uncanny."[232]

If the paranormal explanation is true, most of the known laws of physics are wrong, and the universe is a bizarre place indeed — far stranger than anyone can presently imagine. Of course, there is no such thing as the supernatural, there are only natural laws that are not fully understood. More likely, Bigfoot is either an unknown hominid with an exceptional ability to elude capture or it is a human creation. The evidence for Bigfoot is compelling: gripping eyewitness accounts by credible people; eerie footprints; mysterious hair; strange vocalizations; photos and videos. Ultimately, the truth is out there. Bigfoot either exists or it doesn't. If it does, it seems inevitable that it will be found. If in coming years the mystery continues to linger without any concrete, indisputable proof, this will be the answer in itself. Finding Bigfoot would be the discovery of the millennium and may change our understanding of what it means to be human. Deep down, we all want to believe. The challenge for investigators is this: show us the body. Nothing less will do. In the words of an old English proverb, "Time trieth truth."

In February 2004, the Whitehall Village Board passed legislation protecting Bigfoot from hunting and abuse. Paul Bartholomew proposed and lobbied tirelessly for the ordinance and had it named after Dr. Warren Cook. The following is the exact wording of the new ordinance.

THE DR. WARREN L. COOK SASQUATCH/BIGFOOT PROTECTIVE ORDINANCE

WHEREAS there is an historic traditional history of accumulating reports of a bi-pedal ape-like creature walking like a human in the Whitehall, New York area often referred to as "Bigfoot" or "Sasquatch"

WHEREAS reports of these creatures can be traced back to the Iroquois and Algonquins, and are referenced clearly in the works of Samual de Champlain and represent a consistent pattern of sightings

WHERAS the possibility of all endangered species, proven and pending scientific recognition, should be entitled to protection under both federal and New York State laws

WHEREAS publicity of these creatures could draw not only scientific scrutiny, but unwanted hunting parties with weapons that could pose a lethal threat to both creatures and area residents as well

WHEREAS legislature to protect other cryptozoological creatures has been successfully passed in both Port Henry for The Lake Champlain Monsters popularly referred to as "Champ," and in Skamania County, Washington for "Bigfoot" or "Sasquatch."

NOW THEREFORE, BE IT RESOLVED BY THE VILLAGE BOARD OF WHITEHALL, NEW YORK that the Village of Whitehall, New York adopt the following measures to ensure the safety of those creatures known as "Bigfoot" or "Sasquatch" in the following two sections:

SECTION ONE: SASQUATCH SAFETY PRESERVE. The creatures known as "Sasquatch" or "Bigfoot" are declared an endangered species in the Village of Whitehall, New York and are hereby protected from potentially lethal abuse or annihilation by hunters or hunting parties.

SECTION TWO: CRIME/PUNISHMENT. The willful, premeditated act of killing or fatally injuring a "Sasquatch" or "Bigfoot" within the borders of the Village of Whitehall, New York will be prohibited.

PART 4 **Case Catalog**
Bigfoot-Related Incidents

Chronological Order

Long Ago, NY & VT
Numerous Algonquian-speaking groups inhabiting what is now New York and Vermont states, told stories of the Windigo, said to be a giant cannibalistic man. The creatures resembled many modern-day Bigfoot sightings in the region. The Iroquois had similar legends of half-human creatures with powerful physiques referred to as Stone Giants.

1604, northern NY
French explorer Samuel de Champlain wrote in the log of his voyage on the St. Lawrence River, in what is now Upstate New York, that the native Americans throughout the region feared the Gougou — a large human-like beast said to inhabit the wilderness.

1759, near Missisquoi Bay, northern VT
French and Indian War hero Major Robert Rogers and his famous ranger unit were reportedly harassed by a mysterious bear who tossed pine cones and nuts on them from nearby trees and ledges. Local Indians had a name for the creature — wejuk. The incidents were recounted by a ranger scout for Rogers' Rangers in the historical novel *Northwest Passage* by Kenneth Roberts.

Eighteenth Century, Morgan, Maidstone, Lemington & Victory, VT
There are numerous references to a mischievous, cunning super bear nicknamed Old Slippry Skin (sometimes spelled Slippryskin) who walked on two legs and outwitted the early pioneers.

Early 1800s
Vermont Governor Jonas Galusha led a hunting party in hopes of shooting Slippryskin, but was unsuccessful.

Early 1800s, near Morgan, VT
A group of hunters tracking Slippryskin reported that the "bear" had cleverly backtracked on its prints before sending a large tree rolling down a mountainside toward the party, which then gave up the search.

1818, Ellisburg, Jefferson County, NY
A respected resident reported seeing a hairy wild man running through the woods. It stopped, stared, then ran off. The man said the body of the creature was bending forward as it moved. A massive hunt was launched involving hundreds of residents, but nothing was found.

Summer 1838, near Silver Lake, northern Susquehanna County, NY
A boy was frightened by a mysterious animal that "looked like a human being, covered with black hair, about the size of his brother, who was six or seven years old." The creature made a whistling sound. He grabbed his gun and fired at the animal-man. He missed and the little creature fled into the woods.

August 1861, Bennington, VT & North Adams, MA region
Search parties fired on a "hideous" gorilla-like creature spotted on numerous occasions.

1868, near West Milton, Saratoga County, NY
There were many sightings of a wild man who managed to elude capture.

Mid-June/July 1869, near Woodhill & Troupsville, Steuben County, NY
At least 100 residents reported seeing a wild man in the region. Over 200 men were organized to hunt down the creature. One of the hunters had a clean shot at it but could not shoot; he said that it was too human-like. The creature ran off "with a springing, jerking hitch in his gait." It resembled more "of a wild animal than a human being." A search of the area was fruitless.

August 1869, Ogdensburg, NY
A mysterious man-like creature was seen by several people. The limbs were said to have been "covered with long hair." The wild man reportedly took minnows from a pail and ate them.

Early October 1879, near Williamstown, VT
There was great excitement after two hunters reported seeing a half man, half animal that stood five feet tall and was covered "with bright red hair."

March 1883, Port Henry, NY
The *Plattsburgh Sentinel* reported that the village was in the midst of a wild-man scare. The creature was spotted on numerous occasions and was said to have been covered with fur.

August–November 1883, Maine, NY
A series of hairy man sightings were recorded in this small town northwest of Binghamton. The creature was described as "low in stature, covered with hair, and running while bent close to the ground." Two witnesses made a bizarre observation — that the figure appeared to have no forearms.

November 1888, Big Run, Jefferson County, NY
Huge human-like footprints eighteen inches in length were found in the snow.

Autumn 1893, Rockaway Inlet, Western Long Island, NY
"Red" McDowell and George Farrell were in a boat when they spotted a wild man on shore, uttering "wild cries." Frightened, they paddled away. Other witnesses reported seeing the creature in the area. It was said to be "large in appearance, with fierce, bloodshot eyes" and "long, flowing, matted hair." Theories abounded that it was a missing sailor from the schooner *Maggie Devine,* who went crazy after the vessel ran aground in the area weeks earlier, but there was little evidence to support this.

1894–1895, Norfolk, CT
A gorilla-like creature was spotted by several residents.

August 1895, Winsted, CT
A series of wild-man sightings in the region sparked a massive search for the creature involving over 500 men. Among descriptions: it had a large head, big teeth, was muscular and covered in long, black hair.

Summer 1897, Margaretville near Newburg, NY
Peter Thomas reported that "a wild eyed man or ape" with "long and hairy arms" attacked him and killed one of his horses. He said the entity was "seven feet high, of human shape, covered with hair." The next day, farmer John Cook said he was attacked by "a ferocious ape-like being" that was seven feet tall and covered in black hair.

September 1897, Bronxville, Long Island, NY
Police searched in vain for a wild man with matted hair, seen by numerous residents.

Autumn 1897, Dresden in the Finger Lakes region of Central NY
Hunters scoured the woods for a "strange person or thing" described by a witness who tripped over it as, "a gorilla, being covered with a dark sort of hair or skin."

July 1906, Wakefield near the Bronx, New York City
There were many sightings of a "ferocious looking human covered with more hair than the law permits" and having eyes that glow in the dark.

February 1909, Eastport, Westhampton, Patchogue & Quogue, Long Island, NY
Locals spotted a monkey-like creature "with eyes of flame." Nocturnal search parties were organized and hunters scoured the woods but found no trace of the creature.

July 1909, Haverhill, MA
Police in the city of Haverhill were searching a wooded area near Gile Street for a mysterious "wild man." The woods extended to the border with Newton, New Hampshire.

November 1909, Rainbow Lake near Malone, NY
A long-haired wild man frightened many residents.

September 1915, Ellicottville, Cattaraugus County, Western NY
A wild-man scare broke out that involved a naked person or thing.

September 1921, Towns of Bangor & Brandon, Franklin County, NY
Residents were on an emotional edge following several sightings of a "wild man." Some parents refused to let their children walk home alone from school.

October–November 1921, near Malone, NY
The County sheriff headed up several search parties in hopes of killing or capturing a wild man who was spotted in the region, mostly in the vicinity of Skerry. Some parents even kept their children from school fearing the creature might attack them. A reporter wrote that "lonely females cower behind locked doors and men wag their heads in gossip as they ponder over the puzzle of the wild man..."

Mid-Autumn 1922, near Babylon, NY
Hunters searched the woods for a creature described by some as a baboon, others like a gorilla. The animal was never found.

Summer 1931, Huntington, Long Island, NY
Police searched for a "wandering gorilla or perhaps a chimpanzee" after several sightings. Nassau County Police organized volunteers to search the area; their only clue — strange human-like tracks. On July 18, a creature resembling a gorilla was seen by a family in Huntington crashing through shrubbery.

February 1932, Newcomb, Hamilton County, NY
Two Indian Lake cousins were fur trapping when, near the O'Neil Flow, they spotted a strange hairy creature that left huge footprints. After a police search was mounted, the wild man was found near Dunbrook Mountain and turned out to be a small man wrapped in many layers of fur. The identity of the man, who was shot and killed by State Police, remains a mystery.

August–September 1934, Amityville, Long Island, NY
A search was launched for an ape-like creature. In one instance, it was believed to have been responsible for shredding a fur coat and several mattresses at the Alfred Abernathy residence.

Mid-February 1941, Potsdam, NY
A wild man frightened students near Copeland School in Potsdam, and chased Miss Marion Smith while she walked home from skating.

November 1942, Pontoosuc Lake, MA
After several sightings of a "wild man," an AWOL army officer was arrested on November 18 and was assumed to be responsible for the reports.

Mid-November 1948, near Tupper Lake, NY
Schuylerville hunter Lawrence Peets spotted a wild man making whimpering sounds near deserted Kildare village. The "man" appeared to be covered in rabbit furs and ran into the woods after Peets called out.

February 1951, Town of Sudbury, VT
Lumberjack John Rowell and a co-worker were snaking logs when they noticed that their very heavy oil drum had been moved several hundred feet by something that made giant human footprints in the snow, twenty inches long and eight inches wide.

1959, Whitehall, NY
A Bigfoot-like creature was spotted near Abair Road.

1961, East Bay, Whitehall, NY
Two women were terrified by a large figure that moved towards them as they went for a walk at night. Police searched the area but found nothing.

Early 1960s, Plainfield, VT
Farmer William Lyford spotted a tall, bear-like creature walking on two legs. It ran over a hill when a flashlight was pointed on it.

1964, Stockbridge, Rutland County
Six people traveling in a pick-up truck at 7:30 p.m. spotted a huge, gray creature that stood seven to eight feet tall and walked briskly.

November 1968, South Kent, Litchfield County, CT
A nine-foot tall, hairy creature was spotted from a window near the Newton farm.

Summer 1969, Long Lake, NY
An ape-like creature appeared in the window of a cabin at the Pumphouse campsite. It had a cone-shaped head with brownish fur, and a dark face that had a pushed-in look.

Mid-November 1969, Mountaintop in Southern VT
A deer hunter peered through his rifle scope to see a creature with a huge body that was covered in dark brown hair and holding a white object.

1970, between Glens Falls & Corinth, NY
A large, hairy creature that stood six to seven feet tall, ran in front of a car driven by a schoolteacher. It had a pointed, bumpy head, massive legs and thighs, and was white in color. A passenger confirmed the encounter.

Early 1970s, Galick Nature Preserve, Whitehall, NY
An ape-like creature walked onto a man's porch.

February 16, 1974, Whitehall, NY
A man parking with his girlfriend on the side of the road, was startled by the silhouette of a huge creature towering nearly seven feet tall. He said nothing to his companion, then made up an excuse about having to get back to the village.

Mid-March 1974, Barre, VT
Two men were driving on Country Club Road when they heard a shriek and spotted "a tall, dark figure" with arms hanging below its knees.

September 1974, Wakefield, RI
A university student riding a bicycle on Perry Avenue near the Great Swamp Area spotted a huge gorilla-like creature with white hair, bent knees and massive arms. It had a flat face, deep set eyes and human-like lips. As it moved towards her, she fled on her bike.

December 7, 1974, Watertown, NY
Two boys walking near the St. Andrew's Episcopal Church parking lot spotted a six-foot tall creature that let out a loud roar and had its arms raised into the air. A police search found nothing.

1974, Rutland, VT
A couple observed an eight- to ten-foot tall creature in a field. Police investigating the report, spotted a creature with a similar appearance.

January 21, 1975, Watertown, NY
A large, hairy creature with long arms and standing under six feet tall was spotted by a nurse while driving near a church at 10:45 p.m. She said the figure paid no attention to her as if she were not there. Ten-inch footprints of a four-toed something, were found at the site.

January 22, 1975, Watertown, NY
A bear-like creature was spotted in the parking lot of St. Andrews Episcopal Church.

January 1975, Watertown, NY
Three people saw a five-foot tall creature "just walking and swinging its arms" on State Street Hill.

May 1975, Whitehall, NY
Golf course owner Clifford Sparks encountered a giant, two-legged "sloth-like thing" while watering the greens at 11:30 p.m. It had a cone-shaped head, hardly any neck, and walked clumsily.

June 1975, Saranac Lake, NY
Two men spotted "Bigfoot" squatting on the side of Route 3, before it stood up and disappeared into the bushes.

August–September 1975, Missisquoi Bay, VT
A massive eight-foot-tall creature was seen lumbering through the woods before disappearing. It was dark in color and had long legs.

August 11, 1976, Watertown, NY
Two boys walking on Overland Drive at 5:45 a.m. saw a towering creature in the road. They said it stood eight feet tall, had broad shoulders, and was covered with black hair. It made thumping sounds and let out screams.

August 24, 1976, Whitehall, NY
Martin Paddock, Paul Gosselin and Bart Kinney told police they saw a powerfully built, hairy creature that stood seven to eight-foot tall in a field off Abair Road at 10:00 p.m., sparking a media blitz and searches of the area. Nothing was found.

August 25, 1976, Whitehall, NY
Farmer Harry Diekel found "big, human footprints" in a field near Abair Road. A ravaged deer carcass was nearby.

August 26, 1976, Whitehall, NY
Whitehall police officer Brian Gosselin and an unnamed New York State Trooper encountered a Bigfoot-like creature covered in dark brown hair near midnight off Abair Road. Gosselin said the creature was seven-and-a-half feet tall, weighed 400 pounds, and its face was human-like. Its arms were very long, extending eight to ten inches below its knees, and its eyes bulged half an inch off its face. He took aim with his pistol but couldn't fire as it looked part human. The creature ran off when the Trooper spotted it with his light.

September 1, 1976, Whitehall, NY
Frank McFarren of nearby Granville told police that at 11:10 p.m. he fired his shotgun and rifle at a huge, hairy creature at Carver's Falls Road, not far from Abair Road. A police search turned up nothing.

September 1976, Lewiston, NY
Village police officer Peter Filicetti was picking corn when the rows began to separate after a large animal passed by. Investigating, he

found a 200-yard trail of three-toed footprints. A hunting party searched for the creature but turned up nothing.

Autumn 1976, Whitehall, NY
Whitehall police Sergeant Wilfred Gosselin was hunting near Abair Road when he heard an "eerie high-pitched yell," which sent a herd of cows into a stampede.

Late December 1976, Agawam, MA
The national media focused on this community after twenty-seven-inch-long, human-like footprints were found in a wooded area. A sixteen-year-old boy soon confessed to creating the prints with boards strapped to his feet.

1976, Sawyers Mountain in East Haven, Caledonia County, VT
A woman was sitting near an apple orchard at noon when a large, hairy, muscular ape-like creature hurried past. It stood nearly ten feet tall, had "long arms and appeared to have human-like eyes."

March 1977, South Bay, Whitehall, NY
Royal Bennett and his granddaughter Shannon, using binoculars, spotted what appeared to be a tree stump near Fish Hill Road. Suddenly an amber-colored "thing" about seven-and-a-half feet tall, stood up and walked off, its arms swinging as it moved.

March 1977, Chittenden, VT
A housewife watched a huge, hairy, gorilla-like creature standing in a field near her home. She said it had poor posture and long arms.

May 7, 1977, Hollis, NH
Local police were approached by Gerard St. Louis of Lowell, Massachusetts, who said he and his family were terrorized when a ten-foot-tall, human-like creature began shaking their parked truck as they slept. Police tried to convince the family that they had encountered a bear standing on its hind legs. St. Louis estimated that it weighed 1,000 pounds, and said it certainly wasn't a bear.

July or August 1977, Clarendon, VT
Nancy and John Ingalls spotted a strange creature 6.5 feet tall that

looked part human, part animal, on Route 7 at 10:15 p.m. Oversized naked footprints were found at the site.

Summer 1978, Charleston, Washington County, RI
A mother and son encountered a large yellowish-white "ape" with a broad chest, standing six to seven feet tall. It had a "long, flat face, long massive arms, [and] its head appeared to be without any neck."

September 1978, Town of Porter near Lewiston, NY
Hunters Kevin Mooradian and David Holt found what appeared to be the body of a dead Bigfoot. It turned out to be the body of a bear.

November 28, 1978, Shrewsbury, VT
Huge, human-like foot prints were found in the snow near the home of Mr. and Mrs. David Fretz on Upper Cold River Road.

December 1978, Kinderhook, Columbia County, NY
Grandmother Martha Hallenbeck spotted a "big, black, hairy thing all curled up" on her lawn, and later gigantic human-like footprints in the snow.

1978 near Bridgewater, MA
Joe DeAndrade was standing in the Hockomock Swamp when he spotted a brown, hairy ape-like creature "walking slowly like Frankenstein, into the brush." He founded a local organization to search for the creature but has never seen it again.

Summer 1979, Hamilton County, NY
A hiker watched a juvenile Bigfoot making a thumping noise. The four-and-a-half-foot-tall creature had dark brown fur and made a sound like a woman crying.

Summer 1979, West Haven, VT
A family out fishing were frightened as a huge, hairy creature peered at them over some bushes, then ran off in a hunched over manner. A similar encounter occurred the following week but this time the man had a gun. He could not bring himself to shoot.

December 5, 1979, Kinderhook, NY
While trapping on Cushings Hill, Barry Knights watched four massive, furry creatures walking on two legs and making clacking or grunting sounds.

Late 1970s, Saratoga County, NY
Saratoga County Sheriff's Deputies investigated a disturbance and arrived to find a group of frightened residents who said that something making loud noises had pulled up a tree ten inches in diameter and threw it against a mobile home trailer.

April 1980, Kinderhook, NY
A woman was driving on Route 9 when her car headlights illuminated a 7.5-foot-tall "highly evolved ape," before it disappeared into a wooded area.

June 4, 1980, Lawrenceburg, NY
Camper Fred Renaudo was awakened by the sound of something crashing through the brush. He saw a large white creature. Fifteen-inch, human-like footprints were found nearby.

September 24, 1980, Kinderhook, NY
Martha Hallenbeck and several relatives were terrified by a large creature moving on two legs near Martha's rural home. It "screamed, moaned [and] made guttural noises" before being scared off by shotgun blasts.

November 1980, Kinderhook, NY
Barry Knights and Russell Zbierski were walking near Cushing's Hill when they saw five huge, neckless creatures with cone-shaped heads converge on the road ahead of them.

November 1980, Kinderhook, NY
At the same time as the previous sighting, a woman who lived nearby said a big, hairy creature walked near her house and took food from her trash cans. Her dog went berserk and wet itself.

April 1981, Kinderhook, NY
A woman bicycle-rider saw a huge creature cross Novak Road and vanish into a cornfield.

May 8, 1981, Kinderhook, NY
Campers near Cushing's Hill spotted a tall creature walking on two legs, with no apparent neck and long arms. They said the red eyes seemed to glow in the dark.

Summer 1981, south of Kinderhook, NY
A man encountered a black, hairy Bigfoot near a dead-end road in a remote section of Columbia County. Sixteen-inch footprints were found at the site.

November 1981, Kinderhook, NY
The headlights of a woman's car illuminated a "big two-legged thing, reddish-brown, that ran off into the woods."

February 1982, East Bay, Whitehall, NY
Two village police officers on patrol saw a huge, hairy creature, standing seven-and-a-half feet tall, dash across the road ahead on a frigid winter morning at 4:30 a.m. The creature ambled up a steep slope that led into a mountainous region. Officer Danny Gordon said it moved swiftly and took huge strides, yet it had bad posture and slouched as it moved.

May 1982, Kinderhook, NY
Michael Maab looked across Kinderhook Creek while fishing and saw an eight-foot-tall creature with reddish-brown hair, staring at him from twenty yards away. Maab said it had small eyes and black fingernails. It left after two minutes.

Summer 1982, Kinderhook, NY
An elderly resident described seeing a large, two-legged creature covered with black hair, standing in his yard at dusk.

Winter 1983, near Bridgewater, MA
One cold night, John Baker was trapping muskrats in the Hockomock Swamp when a huge, hairy creature crashed through the

woods and ran into the swamp, passing within a few yards of his canoe.

March 1983, Tinmouth, VT
A middle-aged couple driving in a rural area were stunned to see a huge "giant" moving swiftly over a rocky ridge. "His arms were much longer than a normal man's and he appeared to be much bigger — especially taller — than any man either of us had ever seen."

August 20, 1983, October Mountain, MA
Eric Durant and Frederick Perry of Pittsfield, were camping when they heard strange noises. Near midnight they saw a six- to seven-foot-tall creature with dark brown hair and eerie glowing eyes. Both said it walked on two legs and swung its arms as it moved.

October 7, 1983, near Lake George Village, Warren County, NY
Three bicyclists were riding near French Mountain at 7:45 p.m. when they heard screaming. Their flashlight illuminated two large red eyes that were seven feet off the ground. They fled in fright.

November 1983, near Glens Falls, NY
A deer hunter near Spire Falls Mountain looked on in disbelief as a Bigfoot-like creature jogged past. It appeared to be "slouched" over. The creature ambled up the side of a ridge in a way that no human could.

April 1984, Bellows Falls, Windsor County, VT
James Guyette was delivering newspapers at 5:30 a.m. when he was startled by a huge hair-covered "animal man" on Route 91 near the Hartland Dam. Its long arms swung back and forth as the "thing" walked off.

Late May or early June 1984, Hubbardton, Rutland County, VT
Bruce Bateau was awakened at 3:30 a.m. by a high-pitched shriek so loud that it nearly sounded like a whistle. The next day, a trail of huge, naked footprints was found nearby. He noticed a heavy musty/musky odor, which gave him an eerie feeling.

August 20, 1984, Whitehall, NY
A man observed a seven- to eight-foot tall creature twenty yards away at 10:45 p.m. It quickly disappeared.

August 28, 1984, Whitehall, NY
The same witness from the previous entry spotted a 400-pound Big-foot creature walking near a house at 8:30 p.m. It watched him then ran away.

August 1984, Postenkill, NY
A couple driving on a rural road observed a huge, reddish-blond Bigfoot running in a nearby field, before it vaulted a fence.

November 1984, Colchester, VT
A family of four spotted a huge, towering animal-man in a snow squall while traveling on Route 2 near midnight. It was at least ten feet tall, had very long arms, long, white fur with dirty yellow streaks and memorable amber-yellow eyes.

1984, Caledonia & Essex Counties, VT
Logging magnate Hugo Meyer relayed sightings of Bigfoot-like creatures in northern Vermont.

1984, Chittenden, VT
A local hunter was awakened by loud screams in his yard, then heard a powerful "something" rip the hinges off his cellar door. A foot and hand print were found near the site.

March 4, 1985, Clarendon Flats, Rutland County, VT
Dorothy Mason and son Jeff saw huge, human-like footprints sixteen inches by five inches, trailing near their house and into mountains in the direction of West Rutland.

June 1985, Foster Pond, Peacham, Caledonia County, VT
Two fishermen saw what looked to be one or two bears positioned side by side, on the shore. The mass suddenly stood up and a huge hairy creature walked, then ran on two legs, into the woods.

September 20, 1985, West Rutland, VT
At the home of Ed and Theresa Davis on old Route 4A, Bob Davis and Frank "Fron" Grabowski III spotted a "gorilla-like" creature walking towards them on a dirt road. After tossing stones at them, it ran off. Al Davis doubled back to catch the "prankster" but instead saw it and realized it was too big for a prank. Several large, oversized human tracks were found, pressed into the hard gravel. Anthropologist Warren Cook estimated that the indentations were made by something weighing at least 400 pounds.

November 20, 1985, West Rutland, VT
Two boys playing near the Davis home saw a Bigfoot and fled after one of the youths emptied his BB-gun in its direction.

November 21, 1985, West Rutland, VT
A West Rutland school bus turned in a parking area near the Davis home. At this point, several students including Gary and Donna St. Lawrence, spotted a large, black, hairy creature in a field.

October 1, 1986, Between Castleton & West Rutland, VT
John Bradt, Kerry Bilda and George Dietrich were traveling along Route 4A when their vehicle nearly hit a six- to seven-foot tall creature with collie-length hair, high cheekbones and deep set eyes. It made no attempt to get out of the way.

October 26, 1986, Between Castleton Corners and Poultney, VT
Jill Cortwright was driving on Route 30 when she and passenger Cathy Quill spotted a bear-like creature on the side of the road. They said it appeared to be oblivious to their presence. The hair was "messy, not smooth like a bear."

1987, near Rutland, VT
Farmer John Miller spotted a ten-foot-tall, black creature that resembled a gorilla.

1987, Salisbury, NH
A Webster pheasant hunter spotted a nine-foot-tall Bigfoot that was covered in gray hair. The hands were three times bigger than a

human. It had long arms and legs and was gorilla-like, "but this here wasn't a gorilla."

Autumn 1988, Rutland County, VT
Two teens encountered a large creature in the woods. It walked briskly on two legs, stood seven feet tall and weighed 300 pounds.

August 18, 1989, Hampton, NY
A young man and his dog encountered a dark Bigfoot while walking along Route 22A at 3:00 p.m. The creature had matted hair. The dog whined and ran off.

August 24, 1989, Hampton, NY
The same witness from the previous entry, was camping with a companion at 2:00 a.m. when he spotted two "glowing red eyes" in the woods six feet high, circling the camp. Human-like prints ten inches by four inches were found nearby.

December 1989, Eden, Lamoille County, VT
Guy Primo and his family noticed huge, naked footprints in the snow, thirteen inches by six inches, trailing into the forest.

January 1990, Ghent, NY
Twenty-inch-long footprints were found in a field trailing through the snow. Days later, similar tracks were found nearby.

Autumn 1990, Whitehall, NY
A man watched a creature "on two legs" run down a bank and into some woods by Hickey Road.

Autumn 1993, Whitehall, NY
A brother and sister were waiting for a school bus near Abair Road when they saw a Bigfoot-like creature hunched over in a field.

October 31, 1994, Whitehall, NY
An Abair Road resident encountered a mysterious creature with dark brown hair, "walking" at 7:10 a.m.

March 14, 1995, Clemons, NY
Darrin Gosselin and a friend found a long trail of huge, human-like footprints in the snow near an abandoned graphite mine.

June 1995, Olmsteadville, Essex County, NY
A strange squatting creature was spotted along Route 28 at 12:25 a.m. What may have been a juvenile Bigfoot was said to be five feet tall, weighed eighty pounds, and had brown and white matted hair. Its eyes were described as oval-shaped and "gentle."

August 15, 1996, Saranac Lake, NY
Two fishermen on Pine Lake at dusk spotted what looked to be a black bear that suddenly stood up and walked off.

1997 Plattekill, Ulster County, NY
Lembo Lake camper Doug Pridgen made a videotape during a rock concert in an apple orchard. Upon reviewing the video, two ape-like creatures appeared to be moving with ease among the trees in the background.

February 1998, along the Genesee River, Western, NY
Two girls saw a six-and-a-half-foot-tall, grayish-white Bigfoot jump in front of them on a trail, passing to within six feet. It had a powerful physique, long, matted, shaggy hair, and quickly walked off while growling.

August 1998, Caroga Lake, Fulton County, NY
Two men traveling on North Bush Road at 2:00 a.m., saw a seven-and-a-half-foot-tall animal-man with a flat face and arms that swung in an exaggerated manner.

January 1999, near Ludlow, VT
A hunter spotted a Bigfoot-like creature close up. It was covered with reddish-brown fur that was "matted in places" and appeared to be "shedding." It climbed a steep ridge. Large "human-looking footprints" were found nearby.

September 13, 1999 near Corning, Schulyer County, NY
A man driving on a stretch of highway 224, saw a Bigfoot in the

headlights of an oncoming vehicle, heading from a cornfield and into woods.

July 11, 2000, Lyon Falls, Lewis County, NY
A group of girls from a summer camp observed a "dark, hairy, gorilla-like" creature that was "walking on two legs." It was in view for over a minute. A counselor tried unsuccessfully to convince them that it was a bear.

Autumn 2001, Alleghany Reservation near Salamanca, Cattaraugus County, NY
A hunter saw a tall, hairy creature cross over railroad tracks and noticed a strong skunk-like smell.

May 2002, Hillsborough County, southern NH
The witness, delivering newspapers at 4:45 a.m., saw the silhouette of a hairy Bigfoot-like creature that stood eight-and-a-half feet tall and had a cone-shaped head.

Autumn 2002, near Middleport, Niagara County, NY
A man spotting deer watched a large, two-legged creature with long, dark brown hair, seemingly glide over a rough field in tall grass. He said, "I know what I saw that afternoon was remarkable."

June 2003, Comstock, NY
Larry Paap spotted a strange creature with "golden brown hair" about five inches long. It had a six- to eight-inch neck, was visible for thirty seconds, then seemed to vanish before his eyes.

Autumn 2003, Glastenbury, VT
Ray Dufresne was driving on Route 7 at dusk when he saw a "big black thing" some six feet tall, lumbering into the woods near Glastenbury Mountain at the highest part of the road. He said, "It was hairy from the top of his head to the bottom of his feet."

June 20, 2004, Greece, Monroe County, NY
Three men spotted a Bigfoot-like creature on a bike trail at 12:30 a.m. It was over seven feet tall, had gray or tan fur, and walked into the woods on two legs.

June 2004, Clemons, NY
Two Hong Kong visitors were fishing when they spotted a large monkey-like creature with reddish-brown hair and a flat face, wading through the water at a very fast rate. They said it resembled an orangutan.

October 15, 2004, near Napoli, Cattaraugus County, NY
A turkey hunter saw a "thing" up close. "I didn't know what it was. It was on two legs, all brown-red hair all over it and at least six feet tall."

November 2004, Goat Rock near Stoughton, MA
As three hikers reached the summit of Goat Rock, they noticed a bad smell and watched a large, hairy creature dragging a deer carcass. The reaction by one witness: "absolutely stunned."

January 1, 2005, near Troy, VT
Early on the morning of the New Year, a Quebec trucker and his Vermont girlfriend crossed the Canadian border and encountered a Bigfoot on a desolate stretch of Route 243. He said the creature was seven feet tall, weighed nearly 500 pounds and was visible in the truck headlights for several minutes after he stopped to get a better look. It leaped into the air, then gingerly walked away after the woman tooted the horn.

January 2005 near Bristol, CT
A half-human, half-gorilla with dark black hair and "glowing red eyes" was seen standing five-and-a-half feet tall, walking in the snow.

September 1, 2005, Coos County, northern NH
Two deer hunters heard a high-pitched yell before observing a dark Bigfoot between six and seven feet tall "with a long pointy head." Fifteen- to eighteen-inch footprints were found nearby.

October 8, 2005, Ludlow, VT
Three witnesses were startled by a large two-legged creature as they drove near Okemo Mountain. It was eight feet tall, had a cone-shaped head, huge hands, and a heavy build that was "covered in short dark hair." The arms swung back and forth as it moved.

June 10, 2006, Minerva, NY
A motorist saw a large, brown furry figure seven and a half feet tall, lunge into dense woods.

September 3, 2006, Whitehall, NY
At 9:40 p.m., Rich Martin and his family were traveling into Whitehall from nearby Vermont when they saw a six- to seven-foot tall Bigfoot with dark brown hair, standing by Route 4 shortly before entering the village.

September 4, 2006, Whitehall, NY
Four people were driving along Route 4 at 8:45 p.m. when they spotted a mysterious figure standing beside the road. "It had a white face and black hair," and stood six to seven feet tall.

October 2006, East Whitehall, NY
Three teenage girls camping in a rural part of town, spotted a seven-foot-tall creature described as part human. "We all saw it at the same time and we all screamed," said one girl who "froze" while her companions took off.

November 2006, near Eagle Lake, ME
A deer hunter heard a grunting noise and, upon investigating, saw a large, hairy creature towering nearly eight feet tall. It looked at him for twenty seconds before running into the woods.

Summary of New York Encounters

Long Ago, NY
Numerous Algonquian-speaking groups inhabiting what is now New York state, told stories of the Windigo, said to be a "giant cannibalistic man." The creatures resembled many modern-day Bigfoot sightings in the region. The Iroquois had similar legends of half-human creatures with powerful physiques, referred to as Stone Giants.

1604, Northern NY
French explorer Samuel de Champlain wrote in the log of his voyage on the St. Lawrence River in what is now Upstate New York, that the Native Americans throughout the region feared the Gougou, a large human-like beast said to inhabit the wilderness.

1818 Ellisburg, Jefferson County, NY
A respected resident reported seeing a hairy wild man running through the woods. It stopped, stared, then ran off. The man said the body of the creature was bending forward as it moved. A massive hunt was launched involving hundreds of residents, but nothing was found.

Summer 1838, near Silver Lake, northern Susquehanna County, NY
A boy was frightened by a mysterious animal that "looked like a human being, covered with black hair, the size of his brother, who was six or seven years old." The creature made a whistling sound. He grabbed his gun and fired at the little animal-man. He missed and the creature fled into the woods.

1868, West Milton, Saratoga County, NY
There were several sightings of a wild man who managed to elude capture.

Mid-June/July 1869, near Woodhill & Troupsville, Steuben County, NY
At least 100 residents reported spotting a wild man in the region. Over 200 men were organized to hunt down the creature. One of the hunters had a clean shot at it but could not shoot; he stated that it was too human-like. The creature ran off "with a springing, jerking hitch in his gait." It resembled more "of a wild animal than a human being." A search of the area was fruitless.

August 1869, Ogdensburg, NY
A mysterious man-like creature was seen by several people. The limbs were said to have been "covered with long hair." The wild man reportedly took minnows from a pail and ate them.

March 1883, Port Henry, NY
The *Plattsburgh Sentinel* reported that the village was in the midst of a wild-man scare. The creature was spotted on numerous occasions and was said to have been covered with fur.

August–November 1883, Maine, NY
A series of hairy-man sightings were recorded in this small town northwest of Binghamton. The creature was described as "low in stature, covered with hair, and running while bent close to the ground." Two witnesses made a bizarre observation that the figure appeared to have no forearms.

November 1888, Big Run, Jefferson County, NY
Huge human-like footprints measuring eighteen inches in length, were found in the snow.

Autumn 1893, Rockaway Inlet, Western Long Island, NY
"Red" McDowell and George Farrell were in a boat when they spotted a wild man on shore, uttering "wild cries." Frightened, they paddled away. Other witnesses reported seeing the creature in the area. It was said to be "large in appearance, with fierce, bloodshot eyes" and "long, flowing, matted hair." Theories abounded that it was a missing sailor from the schooner *Maggie Devine,* who went crazy after the vessel ran aground in the area weeks earlier, but there was little evidence to support this.

Summer 1897, Margaretville near Newburg, NY
Peter Thomas reported that "a wild eyed man or ape" with "long and hairy arms" attacked him and killed one of his horses. He said the entity was "seven feet high, of human shape, covered with hair." The next day, farmer John Cook said he was attacked by "a ferocious ape-like being" that was seven feet tall and covered in black hair.

September 1897, Bronxville, Long Island, NY
Police searched in vain for a wild man with matted hair, seen by numerous residents.

Autumn 1897, Dresden, Finger Lakes region, Central NY
Hunters scoured the woods for a "strange person or thing" described by a witness who tripped over it as "a gorilla, being covered with a dark sort of hair or skin."

July 1906, Wakefield near the Bronx, New York City
There were many sightings of a "ferocious looking human covered with more hair than the law permits" and having eyes that glowed in the dark.

February 1909, Eastport, Westhampton, Patchogue and Quogue, Long Island, NY
Locals spotted a monkey-like creature "with eyes of flame." Nightly search parties were organized and hunters scoured the woods but found no trace.

November 1909, Rainbow Lake near Malone, NY
A long-haired wild man frightened many residents.

September 1915, Ellicottville, Cattaraugus County, Western NY
A wild-man scare broke out that involved a naked person or thing.

September 1921, Towns of Bangor & Brandon, Franklin County, NY
Residents were on an emotional edge following several sightings of a wild man. Some parents refused to let their children walk home alone from school.

October–November 1921, near Malone, NY
The County sheriff headed up several search parties in hopes of killing or capturing a wild man who was spotted in the region, mostly in the vicinity of Skerry. Some parents even kept their children from school fearing the creature might attack them. A reporter wrote that "lonely females cower behind locked doors and men wag their heads in gossip as they ponder over the puzzle of the wild man…"

Mid-Autumn 1922, near Babylon, NY
Hunters searched the woods for a creature described by some as a baboon, while others said it was more like a gorilla. The animal was never found.

Summer 1931, Huntington, Long Island, NY
Police searched for a "wandering gorilla or perhaps a chimpanzee" after several sightings. Nassau County Police organized volunteers to search the area; their only clue — strange human-like tracks. On July 18, a creature resembling a gorilla was seen by a family in Huntington, crashing through shrubbery.

February 1932, Newcomb, Hamilton County, NY
Two Indian Lake cousins were fur trapping when, near the O'Neil Flow, they spotted a strange hairy creature that left huge footprints. After a police search was mounted, the wild man was found near Dunbrook Mountain and turned out to be a small man wrapped in many layers of fur. The identity of the man, who was shot and killed by State Police, remains a mystery.

August–September 1934, Amityville, Long Island, NY
A search was launched for an ape-like creature. In one instance, it was believed to have been responsible for shredding a fur coat and several mattresses at the Alfred Abernathy residence.

Mid-February 1941, Potsdam, NY
A wild man frightened students near Copeland School in Potsdam, and chased Miss Marion Smith while she walked home from skating.

Mid-November 1948, near Tupper Lake, NY
Schuylerville hunter Lawrence Peets spotted a wild man making

whimpering sounds near deserted Kildare village. The "man" appeared to be covered in rabbit furs and ran into the woods after Peets called out.

1959, Whitehall, NY
A Bigfoot-like creature was spotted near Abair Road.

1961, East Bay, Whitehall, NY
Two women were terrified by a large figure that moved towards them as they went for a walk at night. Police searched the area but found nothing.

Summer 1969, Long Lake, NY
An ape-like creature appeared in the window of a cabin at the Pumphouse campsite. It had a cone-shaped head with brownish fur and a dark face that had a pushed-in look.

1970, between Glens Falls & Corinth, NY
A large, hairy creature that stood six to seven feet tall, ran in front of a car driven by a schoolteacher. It had a pointed, bumpy head, massive legs and thighs, and was white. A passenger confirmed the encounter.

Early 1970s, Galick Nature Preserve, Whitehall, NY
An ape-like creature walked onto a man's porch.

February 16, 1974, Whitehall, NY
A man parking with his girlfriend on the side of the road, was startled by the silhouette of a huge creature towering nearly seven feet tall. He said nothing to his companion, then made up an excuse about having to get back to the village.

December 7, 1974, Watertown, NY
Two boys walking near the St. Andrew's Episcopal Church parking lot spotted a six-foot-tall creature that let out a loud roar and had its arms raised in the air. A police search found nothing.

January 21, 1975, Watertown, NY
A large, hairy creature with long arms and standing under six feet tall,

was spotted by a nurse while driving near a church at 10:45 p.m. She said the creature paid no attention to her as if she were not there. Ten-inch footprints of a four-toed "something," were found at the site.

January 22, 1975, Watertown, NY
A bear-like creature was spotted in the parking lot of St. Andrews Episcopal Church.

January 1975, Watertown, NY
Three people saw a five-foot tall creature "just walking and swinging its arms" on State Street Hill.

May 1975, Whitehall, NY
Golf course owner Clifford Sparks encountered a giant, two-legged "sloth-like thing" while watering the greens at 11:30 p.m. It had a cone-shaped head, hardly any neck, and walked clumsily.

June 1975, Saranac Lake, NY
Two men spotted "Bigfoot" squatting on the side of Route 3, before it stood up and disappeared into the bushes.

August 11, 1976, Watertown, NY
Two boys walking on Overland Drive at 5:45 a.m., saw a towering creature in the road. They said it stood eight feet tall, had broad shoulders, and was covered with black hair. It made thumping sounds and let out screams.

August 24, 1976, Whitehall, NY
Martin Paddock, Paul Gosselin and Bart Kinney told police they saw a powerfully built, hairy creature that stood seven to eight-foot tall in a field off Abair Road at 10:00 p.m., sparking a media blitz and searches of the area. Nothing was found.

August 25, 1976, Whitehall, NY
Farmer Harry Diekel found "big human footprints" in a field near Abair Road. A ravaged deer carcass was nearby.

August 26, 1976, Whitehall, NY
Whitehall Police officer Brian Gosselin and an unnamed New York

State Trooper encountered a Bigfoot-like creature covered in dark brown hair near midnight off Abair Road. Gosselin said the creature was seven-and-a-half feet tall, weighed 400 pounds, and its face was human-like. Its arms were very long, extending eight to ten inches below its knees, and its eyes bulged half an inch off its face. He took aim with his pistol but couldn't fire as it looked part human. The creature ran off when the Trooper spotted it with his light.

September 1, 1976, Whitehall, NY
Frank McFarren of nearby Granville told police that at 11:10 p.m., he fired his shotgun and rifle at a huge hairy creature at Carver's Falls Road, not far from Abair Road. A police search turned up nothing.

September 1976, Lewiston, NY
Village police officer Peter Filicetti was picking corn when the rows began to separate after a large animal passed by. Investigating, he found a 200-yard trail of three-toed footprints. A hunting party searched for the creature but turned up nothing.

Autumn 1976, Whitehall, NY
Whitehall police Sergeant Wilfred Gosselin was hunting near Abair Road when he heard an "eerie, high-pitched yell," which sent a herd of cows into a stampede.

March 1977, South Bay, Whitehall, NY
Royal Bennett and his granddaughter Shannon, using binoculars, spotted what appeared to be a tree stump near Fish Hill Road. Suddenly an amber-colored "thing" seven-and-a-half feet tall, stood up and walked off, its arms swinging as it moved.

September 1978, Town of Porter near Lewiston, NY
Hunters Kevin Mooradian and David Holt found what appeared to be the body of a dead Bigfoot. It turned out to be the body of a bear.

December 1978, Kinderhook, Columbia County, NY
Grandmother Martha Hallenbeck spotted a "big, black, hairy thing all curled up" on her lawn and, later, gigantic human-like footprints in the snow.

Summer 1979, Hamilton County, NY
A hiker watched a juvenile Bigfoot making a thumping noise. The four-and-a-half-foot-tall creature had dark brown fur and made a sound like a woman crying.

December 5, 1979, Kinderhook, NY
While trapping on Cushings Hill, Barry Knight watched four massive, furry creatures walking on two legs and making clacking or grunting sounds.

Late 1970s, Saratoga County, NY
Saratoga County Sheriff's Deputies investigated a disturbance and arrived to find a group of frightened residents who said that something making loud noises had pulled up a tree ten inches in diameter and threw it against a mobile home trailer.

April 1980, Kinderhook, NY
A woman was driving on Route 9 when her car headlights illuminated a seven-and-a-half-foot-tall "highly evolved ape." It disappeared into a wooded area.

June 4, 1980, Lawrenceburg, NY
Camper Fred Renaudo was awakened by the sound of something crashing through the brush. He saw a large white creature. Fifteen-inch, human-like footprints were found nearby.

September 24, 1980, Kinderhook, NY
Martha Hallenbeck and several relatives were terrified by a large creature moving on two legs near Martha's home. It "screamed, moaned, [and] made guttural noises" before being scared off by shotgun blasts.

November 1980, Kinderhook, NY
Barry Knights and Russell Zbierski were walking near Cushing's Hill when they saw five huge, neckless creatures with cone-shaped heads converge on the road ahead of them.

November 1980, Kinderhook, NY
At the time of the previous sighting, a woman who lived nearby said

a big, hairy creature walked near her house and took food from her trash cans. Her dog went berserk and wet itself.

April 1981, Kinderhook, NY
A woman bicycle rider saw a huge creature cross Novak Road and vanish into a cornfield.

May 8, 1981, Kinderhook, NY
Campers near Cushing's Hill spotted a tall creature walking on two legs, with no apparent neck, and long arms. They said the red eyes seemed to glow in the dark.

Summer 1981, south of Kinderhook, NY
A man encountered a black, hairy Bigfoot near a dead-end road in a remote section of Columbia County. Sixteen-inch footprints were found at the site.

November 1981, Kinderhook, NY
The headlights of a woman's car illuminated a "big two-legged thing, reddish-brown, that ran off into the woods."

February 1982, East Bay, Whitehall, NY
Two village police officers on patrol, saw a huge, hairy creature standing seven and a half feet tall, dash across the road ahead on a frigid winter morning at 4:30 a.m. The creature ambled up a steep slope that led into a mountainous region. Officer Danny Gordon said it moved swiftly and took huge strides, yet it had bad posture and slouched as it moved.

May 1982, Kinderhook, NY
Michael Maab looked across Kinderhook Creek while fishing and saw an eight-foot-tall creature with reddish-brown hair, staring at him from twenty yards away. Maab said it had small eyes and black fingernails. It left after two minutes.

Summer 1982, Kinderhook, NY
An elderly resident described seeing a large, two-legged creature covered with black hair, standing in his yard at dusk.

October 7, 1983, near Lake George Village, Warren County, NY
Three bicyclists were riding near French Mountain at 7:45 p.m. when they heard screaming. Their flashlight illuminated two large red eyes that were seven feet off the ground. They fled in fright.

November 1983, near Glens Falls, NY
A deer hunter near Spire Falls Mountain looked on in disbelief as a Bigfoot-like creature jogged past. It appeared to be "slouched" over. The creature ambled up the side of a ridge in a way that no human could.

August 20, 1984, Whitehall, NY
A man observed a seven- to eight-foot tall creature twenty yards away at 10:45 p.m. It quickly disappeared.

August 28, 1984, Whitehall, NY
The same witness from the previous entry, spotted a 400-pound Bigfoot creature walking near a house at 8:30 p.m. It watched him, then ran away.

August 1984, Postenkill, NY
A couple driving on a rural road observed a huge, reddish-blond Bigfoot running in a nearby field, before it vaulted a fence.

August 18, 1989, Hampton, NY
A young man and his dog encountered a dark Bigfoot while walking along Route 22A at 3:00 p.m. The creature had matted hair. The dog whined and ran off.

August 24, 1989, Hampton, NY
The same witness from the previous entry was camping with a companion at 2:00 a.m. when he spotted two "glowing red eyes" in the woods six feet high, circling the camp. Human-like prints ten inches by four inches were found nearby.

January 1990, Ghent, NY
Twenty-inch long footprints were found in a field trailing through the snow. Days later, similar tracks were found in the vicinity.

Autumn 1990, Whitehall, NY
A man watched a creature "on two legs" run down a bank and into a patch of woods by Hickey Road.

Autumn 1993, Whitehall, NY
A brother and sister were waiting for a school bus near Abair Road when they saw a Bigfoot-like creature hunched over in a field.

October 31, 1994, Whitehall, NY
An Abair Road resident encountered a mysterious creature with dark brown hair "walking" at 7:10 a.m.

March 14, 1995, Clemons, NY
Darrin Gosselin and a friend found a long trail of huge, human-like footprints in the snow near an abandoned graphite mine.

June 1995, Olmsteadville, Essex County, NY
A strange squatting creature was spotted along Route 28 at 12:25 a.m. What may have been a juvenile Bigfoot was said to have stood five feet tall, weighed eighty pounds, and had brown and white matted hair. Its eyes were described as oval-shaped and "gentle."

August 15, 1996, Saranac Lake, NY
Two fishermen on Pine Lake at dusk spotted what looked to be a black bear that suddenly stood up and walked off.

1997, Plattekill, Ulster County, NY
Lembo Lake camper Doug Pridgen made a videotape during a rock concert in an apple orchard. Upon reviewing the video, two ape-like creatures appeared to be moving with ease among the trees in the background.

February 1998, along the Genesee River, Western, NY
Two girls saw a six-and-a-half-foot-tall, grayish-white Bigfoot jump in front of them on a walking trail, passing to within six feet. It had a powerful physique, long, matted, shaggy hair, and quickly walked off while growling.

August 1998, Caroga Lake, Fulton County, NY
Two men traveling on North Bush Road at 2:00 a.m., saw a seven-and-a-half-foot-tall animal-man with a flat face and arms that swung in an exaggerated manner.

September 13, 1999 near Corning, Schulyer County, NY
A man driving on a stretch of highway 224, saw a Bigfoot creature in the headlights of an oncoming vehicle, heading from a cornfield and into woods.

July 11, 2000, Lyon Falls, Lewis County, NY
A group of girls from a summer camp observed a "dark, hairy, gorilla-like" creature that was "walking on two legs." It was in view for over a minute. A counselor tried unsuccessfully to convince them that it was a bear.

Autumn 2001, Alleghany Reservation near Salamanca, Cattaraugus County, NY
A hunter saw a tall, hairy creature cross over railroad tracks and noticed a strong skunk-like smell.

Autumn 2002, near Middleport, Niagara County, NY
A man spotting deer watched a large, two-legged creature with long, dark brown hair, seemingly glide over a rough field in tall grass. He said, "I know what I saw that afternoon was remarkable."

June 2003, Comstock, NY
Larry Paap spotted a strange creature with "golden brown hair" about five inches long. It had a six- to eight-inch neck, was visible for thirty seconds, then seemed to vanish before his eyes.

June 20, 2004, Greece, Monroe County, NY
Three men spotted a Bigfoot-like creature on a bike trail at 12:30 a.m. It was over seven feet tall, had gray or tan fur, and walked into the woods on two legs.

June 2004, Clemons, NY
Two Hong Kong visitors were fishing when they spotted a large monkey-like creature with reddish-brown hair and a flat face, wad-

ing through the water at a very fast rate. They said it resembled an orangutan.

October 15, 2004, near Napoli, Cattaraugus County, NY
A turkey hunter saw a "thing" up close. "I didn't know what it was. It was on two legs, all brown-red hair all over it and at least six feet tall."

June 10, 2006, Minerva, NY
A motorist saw a large, brown, furry figure seven-and-a-half feet tall, lunge into dense woods.

September 3, 2006, Whitehall, NY
At 9:40 p.m., Rich Martin and his family were traveling into Whitehall from nearby Vermont when they saw a six- to seven-foot-tall Bigfoot with dark brown hair, standing by Route 4 shortly before entering the village.

September 4, 2006, Whitehall, NY
Four people were driving along Route 4 at 8:45 p.m., when they spotted a mysterious figure standing beside the road. "It had a white face and black hair" and stood six to seven feet tall.

October 2006, East Whitehall, NY
Three teenage girls camping in a rural part of town, spotted a seven-foot-tall creature described as part human. "We all saw it at the same time and we all screamed," said one girl, who "froze" while her companions took off.

Summary of New England Encounters

Connecticut

1894–1895, Norfolk, CT
A gorilla-like creature was spotted by several residents.

August 1895, Winsted, CT
A series of wild-man sightings in the region sparked a massive search for the creature involving over 500 men. Among descriptions: it had a large head, big teeth, was muscular and covered in long, black hair.

November 1968, South Kent, Litchfield County, CT
A nine-foot-tall, hairy creature was spotted from a window near the Newton farm.

January 2005, near Bristol, CT
A half-human, half-gorilla with dark black hair and "glowing red eyes" was seen walking in the snow. It stood five-and-a-half feet tall.

Maine

November 2006, near Eagle Lake, ME
A deer hunter heard a grunting noise and, upon investigating, saw a large, hairy creature towering nearly eight feet tall. It looked at him for twenty seconds before running into the woods.

Massachusetts

August 1861, North Adams, MA region
Search parties fired on a gorilla-like creature that was seen on several occasions.

July 1909, Haverhill, MA
Police in the city of Haverhill were searching a wooded area near Gile Street for a mysterious wild man. The woods extended to the border with Newton, New Hampshire.

November 1942, Pontoosuc Lake, MA
After several sightings of a wild man, an AWOL army officer was arrested on November 18 and was assumed to be responsible for the reports.

Late December 1976, Agawam, MA
The national media focused on this community after twenty-seven-inch-long, human-like footprints were found in a wooded area. A sixteen-year-old boy later confessed to creating the prints with boards strapped to his feet.

1978, near Bridgewater, MA
Joe DeAndrade was standing in the Hockomock Swamp when he spotted a brown, hairy ape-like creature "walking slowing like Frankenstein, into the brush." He founded a local organization to search for the creature but has never seen it again.

Winter 1983, near Bridgewater, MA
One cold night, John Baker was trapping for muskrat pelts in the Hockomock Swamp when a huge, hairy creature crashed through the woods and ran into the swamp, passing within a few yards of his canoe.

August 20, 1983, October Mountain, MA
Eric Durant and Frederick Perry of Pittsfield, were camping when they heard strange noises. Near midnight they saw a six- to seven-foot tall creature with dark brown hair and eerie glowing eyes. Both said it walked on two legs and swung its arms as it moved.

November 2004, Goat Rock near Stoughton, MA
As three hikers reached the summit of Goat Rock, they noticed a bad smell and watched a large hairy creature dragging a deer carcass. The reaction by one witness: "absolutely stunned."

New Hampshire

May 7, 1977, Hollis, NH
Local police were approached by Gerard St. Louis of Lowell, Massachusetts, who said he and his family were terrorized when a ten-foot-tall, human-like creature began shaking their parked truck as they slept. Police tried to convince the family that they had encountered a bear standing on its hind legs. St. Louis estimated that it weighed 1,000 pounds, and said it certainly was not a bear.

1987, Salisbury, NH
A Webster pheasant hunter spotted a nine-foot-tall, Bigfoot-like creature that was covered in gray hair. The hands were three times bigger than a human. It had long arms and legs and was gorilla-like, "but this here wasn't a gorilla."

May 2002, Hillsborough County, southern NH
The witness, delivering newspapers at 4:45 a.m., saw the silhouette of a hairy creature that stood eight-and-a-half feet tall and had a cone-shaped head.

September 1, 2005, Coos County, northern NH
Two deer hunters heard a high-pitched yell before observing a dark Bigfoot between six and seven feet tall "with a long pointy head." Fifteen- to eighteen-inch footprints were found nearby.

Rhode Island

September 1974, Wakefield, RI
A university student riding a bicycle on Perry Avenue near the Great Swamp Area, spotted a huge, gorilla-like creature with white hair, bent knees and "massive arms." It had a flat face, deep set eyes and human-like lips. As it moved towards her, she fled on her bike.

Summer 1978, Charleston, Washington County, RI
A mother and son encountered a large yellowish-white "ape" with a broad chest, standing six to seven feet tall. It had a "long, flat face, long massive arms, [and] its head appeared to be without any neck."

Vermont

Long Ago, VT

There were numerous accounts of Bigfoot-like creatures among the Algonquian-speaking groups inhabiting what is now Vermont, who referred to them as Stone Giants.

1759 near Missisquoi Bay, northern VT

French and Indian war hero Major Robert Rogers and his famous ranger unit were reportedly harassed by a mysterious bear who tossed pine cones and nuts on them from nearby trees and ledges. Local Indians had a name for the creature — wejuk. The incidents were recounted by a ranger scout for Rogers' Rangers in the historical novel *Northwest Passage* by Kenneth Roberts.

Eighteenth Century, Morgan, Maidstone, Lemington & Victory, VT

There are numerous references to a mischievous, cunning super bear nicknamed Old Slippry Skin (sometimes spelled Slippryskin) who walked on two legs and outwitted the early pioneers.

Early 1800s, VT

Vermont Governor Jonas Galusha led a hunting party in hopes of shooting Slippryskin, but was unsuccessful.

Early 1800s near Morgan, VT

A group of hunters tracking Slippryskin reported that the "bear" had cleverly backtracked on its prints before sending a large tree rolling down a mountainside toward the party, which then gave up the search.

August 1861, Bennington, VT

Search parties took shots at a "hideous" gorilla-like creature that was spotted on numerous occasions.

Early October 1879, near Williamstown, VT

There was great excitement after two hunters reported seeing a half man, half animal that stood five feet tall and was covered "with bright red hair."

February 1951, Town of Sudbury, VT
Lumberjack John Rowell and a co-worker were snaking logs when they noticed that their very heavy oil drum had been moved several hundred feet by something that made giant human footprints in the snow, twenty inches long and eight inches wide.

Early 1960s, Plainfield, VT
Farmer William Lyford spotted a tall, bear-like creature walking on two legs. It ran over a hill when a flashlight was pointed on it.

1964, Stockbridge, Rutland County
Six people traveling in a pick-up truck at 7:30 p.m. spotted a huge, gray creature that stood seven to eight feet tall and walked briskly.

Mid-November 1969, Mountaintop in Southern VT
A deer hunter peered through his rifle scope to see a creature with a huge body that was covered in dark brown hair and holding a white object.

Mid-March 1974, Barre, VT
Two men were driving on Country Club Road when they heard a shriek and spotted "a tall, dark figure" with arms hanging below its knees.

1974, Rutland, VT
A couple observed an eight- to ten-foot tall creature in a field. Police investigating the report, spotted a creature with a similar appearance.

August–September 1975, Missisquoi Bay, VT
A massive eight-foot-tall creature was seen lumbering through the woods before disappearing. It was dark in color and had long legs.

1976, Sawyers Mountain, East Haven, Caledonia County, VT
A woman was sitting near an apple orchard at noon when a large, hairy, muscular, ape-like creature hurried past. It stood nearly ten feet tall, had "long arms and appeared to have human-like eyes."

March 1977, Chittenden, VT
A housewife watched a huge, hairy, gorilla-like creature standing in a field near her home. She said it had poor posture and long arms.

July or August 1977, Clarendon, VT
Nancy and John Ingalls spotted a strange creature that was six-and-a-half feet tall and looked part human, part animal, on the side of Route 7 at 10:15 p.m. Oversized, naked footprints were found at the site.

November 28, 1978, Shrewsbury, VT
Huge, human-like footprints were found in the snow near the home of Mr. and Mrs. David Fretz on Upper Cold River Road.

Summer 1979, West Haven, VT
A family out fishing were frightened as a huge, hairy creature peered at them over some bushes, then ran off in a hunched over manner. A similar encounter occurred the following week, but this time the man had a gun. He could not bring himself to shoot.

March 1983, Tinmouth, VT
A middle-aged couple driving in a rural area were stunned to see a huge "giant" moving swiftly over a rocky ridge. "His arms were much longer than a normal man's and he appeared to be much bigger — especially taller — than any man either of us had ever seen."

April 1984, Bellows Falls, Windsor County, VT
James Guyette was delivering newspapers at 5:30 a.m. when he was startled by a huge, hair-covered "animal man" on Route 91 near the Hartland Dam. Its long arms swung back and forth as the "thing" walked off.

Late May or early June 1984, Hubbardton, Rutland County, VT
Bruce Bateau was awakened at 3:30 a.m. by a high-pitched shriek so loud that it nearly sounded like a whistle. The next day, a trail of huge, naked footprints was found nearby. He noticed a heavy musty/musky odor which gave him an eerie feeling.

November 1984, Colchester, VT
A family of four spotted a huge, towering animal-man in a snow squall while traveling on Route 2, near midnight. It was at least ten feet tall, had very long arms, long, white fur with dirty yellow streaks and memorable amber-yellow eyes.

1984, Caledonia & Essex Counties, VT
Logging magnate Hugo Meyer relayed sightings of Bigfoot-like creatures in northern Vermont.

1984, Chittenden, VT
A local hunter was awakened by loud screams in his yard, then heard a powerful "something" rip the hinges off his cellar door. A foot and hand print were found near the site.

March 4, 1985, Clarendon Flats, Rutland County, VT
Dorothy Mason and son Jeff saw huge, human-like footprints sixteen inches by five inches, trailing away from their house and into mountains in the direction of West Rutland.

June 1985, Foster Pond, Peacham, Caledonia County, VT
Two fishermen saw what looked to be one or two bears positioned side by side, on shore. The mass suddenly stood up and a huge, hairy creature walked, then ran, on two legs into the woods.

September 20, 1985, West Rutland, VT
At the home of Ed and Theresa Davis on old Route 4A, Bob Davis and Frank "Fron" Grabowski III spotted a "gorilla-like" creature walking towards them on a dirt road. After tossing stones at them, it ran off. Al Davis doubled back to catch the "prankster" but instead saw it and realized it was too big for a prank. Several large, oversized human tracks were found, pressed into the hard gravel. Anthropologist Warren Cook estimated that the indentations were made by something weighing at least 400 pounds.

November 20, 1985, West Rutland, VT
Two boys playing near the Davis home saw a Bigfoot and fled after one of them emptied his BB-gun in its direction.

November 21, 1985, West Rutland, VT
A West Rutland school bus turned in a parking area near the Davis home. At this point, several students including Gary and Donna St. Lawrence, spotted a large, black, hairy creature in a field.

October 1, 1986, Between Castleton and West Rutland, VT
John Bradt, Kerry Bilda and George Dietrich were traveling along Route 4A when their vehicle nearly hit a six- to seven-foot tall creature with collie-length hair, high cheekbones and deep-set eyes. It made no attempt to get out of the way.

October 26, 1986, Between Castleton Corners and Poultney, VT
Jill Cortwright was driving on Route 30 when she and passenger Cathy Quill spotted a bear-like creature on the side of the road. They said it appeared to be oblivious to their presence. The hair was "messy, not smooth like a bear."

1987, near Rutland, VT
Farmer John Miller spotted a ten-foot-tall, black creature that resembled a gorilla.

Autumn 1988, Rutland County, VT
Two teens encountered a large creature in the woods. It walked briskly on two legs, stood seven feet tall and weighed 300 pounds.

December 1989, Eden, Lamoille County, VT
Guy Primo and his family noticed huge, naked footprints in the snow, thirteen inches by six inches, trailing into the forest.

January 1999, near Ludlow, VT
A hunter spotted a Bigfoot-like creature close up. It was covered with reddish-brown fur that was "matted in places" and appeared to be "shedding." It climbed a steep ridge. Large "human-looking footprints" were found nearby.

Autumn 2003, Glastenbury, VT
Ray Dufresne was driving on Route 7 at dusk when he saw a "big, black thing" some six feet tall, lumbering into the woods near Glas-

tenbury Mountain at the highest part of the road. He said, "It was hairy from the top of his head to the bottom of his feet."

January 1, 2005 near Troy, VT

Early on the morning of the New Year, a Quebec trucker and his Vermont girlfriend crossed the Canadian border and encountered a Bigfoot on a desolate stretch of Route 243. He said the creature was seven feet tall, weighed nearly 500 pounds and was visible in the truck headlights for several minutes after he stopped to get a better look. It leaped into the air, then gingerly walked away after the woman tooted the horn.

October 8, 2005, Ludlow, VT

Three witnesses were startled by a large two-legged creature as they drove near Okemo Mountain. It was eight feet tall, had a cone-shaped head, huge hands, and a heavy build that was "covered in short dark hair." The arms swung back and forth as it moved.

End Notes

1. Tripp, Rhoda Thomas (compiler). *The International Thesaurus of Quotations.* New York: Thomas Y. Crowell, p. 669.

2. Heiman, Michael K. "Adirondack Mountains," Discovery Channel School, original content provided by World Book online, http://www.discoveryschool.com/students/homeworkhelp/worldbook/atozgeography/a/004680.html, June 13, 2002.

3. Guinard, Joseph E., (1930). "Witiko Among the Tete-de-Boule," in *Primitive Man,* Volume 3, pp. 69–70. See also, Rapp, Marvin A. (1956). "Legend of the Stone Giants," *New York Folklore Quarterly* 12(4):280–282; Clark, Jerome and Coleman, Loren (1978). *Creatures of the Outer Edge.* New York: Warner, 1978.

4. Podruchny, Carolyn (2004). "Werewolves and Windigos: Narratives of Cannibal Monsters in French-Canadian Voyageur Oral Tradition." *Ethnohistory* 51(4):677–700. See p. 691; Emerson, Ellen R. (1884). *Indian Myths Illustrated or Legends, Traditions, and Symbols of the Aborigines of America.* Boston, Massachusetts: James R. Osgood and Company, p. 436.

5. Krueger, William Kent (2004). "The Windigo." *Boy's Life* 94(3):40–44. This particular contemporary story is from the Ojibwe, who are part of the Algonquian language group.

6. Ibid.

7. Pitcher, John (2001). "Close Encounters." *Rochester Democrat and Chronicle,* October 28, pages 1C, 10C.

8. Rapp (1956). op cit., pages 280–281.

9. Cooper, John M., *Primitive Man,* Volume 6, "The Cree Witiko Psychosis," pp. 20–24. See also: Bruchac, Joseph (1979). *Stone Giants & Flying Heads. Adventure Stories of the Iroquois.* Crossing Press; (1892). "Iroquois Notes." *The Journal of American Folklore* 5(18) (July–September):223–229.

10. These excerpts are taken from an English translation from the logs of Samuel de Champlain of the experiences he recorded while exploring the St. Lawrence Valley in 1604. They appeared in Volume 1 of a six-volume set, *The Works of Samuel de Champlain, reprinted, translated and annotated by six Canadian scholars under the general editorship of H.P. Biggar.*

Volume 1 covers 1599–1607, translated and edited by H.H. Langton and W.F. Ganong, with the French texts collated by J. Home Cameron. Reprinted by the University of Toronto Press, 1971. The narration appears on pages 186–188 in Chapter XII, titled, "Of a frightful monster which the savages call Gougou, and of our short and favourable passage back to France."

11. "Another Wonder." *Exeter Watchman,* September 22, 1818.

12. *Dorchester Aurora* (Maryland), August 27, 1838.

13. *Utica Morning Herald and Daily Gazette* (Utica, New York), 1868. The exact date is unreadable on the microfilm.

14. "A Wild Man—A Hideous Monster Roaming About the Neighborhood of Woodhill and Troupsburgh, N.Y." *The Evening Gazette,* July 10, 1869. One witness said the creature appeared to be wearing pants, which would indicate that it was a hermit, though it is difficult to draw conclusions from one report, as others gave very different descriptions indicating that it was not a human.

15. "A Wild Man in the Woods." *Plattsburgh Sentinel,* August 6, 1869, p. 1.

16. No title. *Plattsburgh Sentinel,* March 9, 1883, p. 1.

17. "A Mysterious Creature." *The Free Press* (Gouverneur, New York), November 28, 1883.

18. No title. *The Olean Democrat,* (Olean, New York), November 8, 1888.

19. "A Supposed Wild Man." *Democrat Chronicle* (Rochester, New York), December 3, 1883, p. 1. See also, "Hunting a Wild Man." *Buffalo Express,* December 3, 1885, p. 1.

20. *Newburg Daily Press* (New York), July 29, 1895, "Delaware County's Wild Man." Also, *The New York Herald,* July 31, 1895, "He, She or It—Beast or Human. Wild Thing Loose in Delaware County and Scaring the Natives Half Silly."

21. *Illustrated Buffalo Express,* August 4, 1895. Exact page number unreadable.

22. *New York Herald,* November 29, 1893.

23. "Hunt for a Wild man." *Hornellsville Weekly Tribune,* September 3, 1897, p. 1.

24. "Crazy Man or Gorilla." *Oswego Daily Times,* May 17, 1899, p. 6.

25. "Wild Man at Large." *Rochester Democrat and Chronicle,* April 15, 1899, p. 4.

26. "A 'Wild Man' in the Bronx." *The Kansas City Star,* July 28, 1906, p. 3.

27. "Shrieking Apparition Rouses Long Island: Wild, Weird Cries Disturb the Thickets in the Neighborhood of Quogue, and Armed Men, All in Vain, Seek the Lair of the Mysterious Thing." *New York Herald,* February 7, 1909, Section 2, p. 7.

28. "Wild Man Seen in Woods near Rainbow Lake." *The Post-Standard* (Syracuse, New York), November 18, 1909, p. 12; "Wild Man at Rainbow Lake." *Ticonderoga Sentinel,* November 25, 1909, p. 1; "Wild Man of Rainbow Lake," *The Ogdensburg News,* November 19, 1909, p. 1; "Wild Man in the Woods near Rainbow Lake," *Utica Herald Dispatch*, November 18, 1909, p. 5; "Wild Man Running Loose in Adirondacks." *Watertown Daily Times,* November 18, 1909, p. 5.

29. "Wild Man Roaming in Woods near Ellicottville, Startled Villagers Say." *The Illustrated Buffalo Express,* September 26, 1915, p. 1.

30. "Northern New York Items." *Plattsburgh Sentinel,* September 13, 1921, p. 2.

31. "Is Terrorized by Wild Man. Adirondack Community in Veritable Reign of Terror Over Mystery Man." *Dunkirk Evening Observer,* November 5, 1921, p. 14.

32. "Just a Bootlegger's Ruse." *Lake Placid News,* September 21, 1921, p. 16.

33. *Washington Post (*Washington, D.C.), November 6, 1922, p. 1, "Seek Ferocious Baboon in Wilds of Long Island."

34. Keel, John A., *Strange Creatures from Times and Space.* Greenwich, CT: Fawcett, 1970, pp. 95–96.

35. Numerous sources were used to compile this account: "Armed Negro Shot Down in Woods Fight. Troopers Kill Him After Spectacular Pursuit in Mountains." *Syracuse Herald,* March 5, 1932, p. 6; "Negro Wild Man Slain by Posse in Adirondacks." *Oswego–Palladium Times,* March 5, 1932, p. 1; Peasley, L. (1941). "Letter to the Editor." *Ticonderoga Sentinel,* November 13, 1941, pp. 1, 4.

36. Peasley, L. (1941). "Letter to the Editor." *Ticonderoga Sentinel,* November 13, pp. 1, 4; "The Negro Shooting (Editorial)." *Ticonderoga Sentinel,* March 31, 1932, p. 2; "Death Blocks Any Solution of Man's Life." *The Adirondack Record–Elizabethtown Post* (Au Sable Forks, New York), March 24, 1932, p. 1.

37. Ansen, Jay (1977). *The Amityville Horror.* New York: Bantam.

38. "Man, Beast or Demon? It's Loose in Amityville. Mysterious Ape-like Marauder Raids Garage; Town on Guard." *New York Herald Tribune,* September 4, 1934.

39. "Report Wild Man in Copeland School Section." *Commercial Advertiser* (Canton, New York), February 18, 1941, p. 4.

40. "Hunters Claim they saw 'Wild Man' in Adirondacks." *The Evening Observer* (Dunkirk, New York), November 17, 1948, p. 11.

41. Anonymous (1976). "Officers Track Creature." *The Post–Star* (Glens Falls, New York), August 30, p. 1.

42. Interview with one of the witnesses and Paul Bartholomew.

43. Report #1524, submitted by the witness (J.G.) to the Bigfoot Research Organization on March 3, 2000, and a follow-up interview by Bill P. of the BFRO.

44. Interview between Susan and Paul Bartholomew.

45. Paul Bartholomew interview with the man.

46. Paul Bartholomew interview with the witness.

47. Green, John (1978). *Sasquatch: The Apes Among Us.* Seattle, Washington: Hancock House, 1978, pp. 233–234, citing the *Watertown Daily Times,* August 2, 1977. An interview with the witnesses was conducted by Milton LaSalle of Watertown.

48. Modern, Thomas, "Trashquatch: The Hunt for Staten Island's Bigfoot." *New York Press,* accessed December 24, 2006 at: http://www.nypress.com/16/ 18/news&columns/feature.cfm; Green, John (1978). op cit., p. 233.

49. Interview between Cliff Sparks and Bob Bartholomew, Fall 1984, and subsequent interviews by Paul Bartholomew.

50. Bord, Janet, and Bord, Colin (1982). op cit.

51. Green, John (1978). op cit., p. 234. Based on an interview between Milton LaSalle and the witnesses.

52. Audio recording of an on-the-scene report by television journalist.

53. Tracy Egan for WRGB, Channel 6, Albany, New York, circa August 30, 1976.

54. Tracy Egan report (1976). op cit. Anonymous (1976). "Officers Track Creature." *The Post–Star* (Glens Falls, New York), August 30.

55. Tracy Egan report (1976). op cit. Anonymous (1976). "Officers Track

Creature." op cit. Interview between Bill Brann and the three witnesses, in Bartholomew, Paul and Bob, Bruce Hallenbeck and Bill Brann (1991). *Monsters of the Northwoods*. Rutland, VT: the authors, pp. 17–18.

56. Bartholomew et al. (1991). op cit., pp. 17–18, based on an interview by Bill Brann with Paul Gosselin and Bart Kinney.

57. Lewis, Ron (1985b). Video tape of the second "Sasquatch, alias Bigfoot Conference" held at Castleton State College in Castleton, Vermont on April 24, 1985.

58. Bill Brann interview with Brian Gosselin soon after the incident, published in Hallenbeck, Bruce and Bob and Paul Bartholomew (1983). "Bigfoot in the Adirondacks." *Adirondack Bits 'n Pieces Magazine* 1(3) (Spring–Summer), p. 26.

59. Bartholomew et al. (1991). op cit., p. 19, based on an interview by Bill Brann with Brian Gosselin.

60. Based on an interview with Wilfred Gosselin conducted by Bob and Paul Bartholomew and Bill Brann.

61. Interview between Paul Bartholomew and Whitehall Police Dispatcher Robert Martell. Also, interview between Bob Bartholomew and Whitehall Police Dispatcher Fred Palmer. The police log is not entirely accurate. Gun historian Michael Pluta notes that a 12-guage shotgun firing either slugs or shot is not a "caliber" and therefore not designated with the decimal prior to the number.

62. Powless, Linda (1980). "The Lewiston Bigfoot: A Big Hoax Or A Mutant? Scary Creatures Terrorize Lewiston Residents." *Niagara Falls Review,* March 17, p. 4.

63. Billington, Michael, "Bigfoot? Shadowy Creature Keeps to Dark Places in Lewiston," *Buffalo Courier–Express,* March 3, 1980, p. 4.

64. Bartholomew et al. (1991). op cit., p. 29, based on interviews between the witnesses and Bill Brann.

65. Bartholomew et al. (1991). op cit., p. 29, Bill Brann interviews with the witnesses.

66. "News Exclusive...Bigfoot Photos Baffle Experts..." *Weekly World News,* April 29, 1980, pp. 1, 19.

67. Letter from Dr. Sidney Anderson, American Museum of Natural History, Central Park West at 79th St., New York City 10024, dated May 26, 1980, to Paul Bartholomew; Letter from Richard W. Thorington, Curator

of Mammals, National Museum of Natural History, Smithsonian Institution, Washington, D.C., dated April 30, 1980, to Paul Bartholomew.

68. Powless, Linda (1980). op cit.

69. Bartholomew, et al. (1991). op cit., pp. 30–31.

70. Interviews with the witness by Bruce Hallenbeck and Paul Bartholomew.

71. Report #6695, submitted to the Bigfoot Research Organization on August 3, 2003.

72. Clark, Curtis (1984). "Local Man Spots Bigfoot in NY State—An Encounter With a Yeti," *Newton Bee* (Connecticut), June 22, pp. 19–20.

73. Paul and Bob Bartholomew, numerous personal interviews with Bruce and Martha Hallenbeck.

74. Interview with the witness by Bruce Hallenbeck.

75. Interview with the witness and Bruce Hallenbeck.

76. Based on interviews between the officers and Paul Bartholomew. The officers drove Paul to the site on more than one occasion to reconstruct the incident.

77. Interviews between Daniel Gordon and Paul Bartholomew, conducted on numerous occasions since 1982.

78. Bruce Hallenbeck interview with the witness.

79. Bruce Hallenbeck interview with the witness.

80. Bartholomew, et al. (1991). op cit., p. 34, based on Bill Brann interviews.

81. Phone interview between the man and Paul Bartholomew on January 29, 1992.

82. Paul Bartholomew interview with the family.

83. A 1984 letter from the couple sent to Paul Bartholomew.

84. Bartholomew, Paul B. (1989). *The Whitehall Times,* September 7, 1989, p. 1.

85. The article was based on interviews by Paul Bartholomew and Dr. Warren Cook.

86. From interviews conducted by Bruce Hallenbeck.

87. Paul Bartholomew interview with the witness.

88. Paul Bartholomew interview conducted in October 1994.

89. Interview between Paul Bartholomew and the witness.

90. Paul Bartholomew interview with the witnesses.

91. The quotations are based on a report submitted by one of the witnesses, to the Bigfoot Research Organization, filed as #10289 and submitted January 18, 2005. BFRO investigator Paul J. Mateja conducted a follow-up with the witness.

92. Horrigan, Jeremiah (2005). "Seeing Sasquatch," *Times Herald-Record* (New York), November 24.

93. Based on an interview by Paul J. Mateja for the Bigfoot Research Organization who interviewed the witness on June 23, 2005, and filed it as report #11726. The report also includes a statement submitted by the witness on May 16, 2005.

94. Report #4286 submitted by "Chris" to the Bigfoot Research Organization on May 10, 2002, and a follow-up interview by an unnamed BFRO investigator.

95. Bruun, Christine, "Bigfoot Near the Finger Lakes? – New York State," Sasquatch Watch, accessed on April 15, 2007 at: http://www.cactusventures.com/webstuff6/finger_lakes_bigfoot.htm.

96. We have chosen to omit several seemingly strange reports recorded between 1966 and 1969, involving a series of bizarre Bigfoot accounts coming out of Western Long Island, often associated with UFO sightings. As improbable as it sounds, numerous couples parking in the vicinity of Mount Misery, in Huntington, Long Island, expecting a quiet, romantic night, reported seeing more than they bargained for — a hairy, seven-foot-tall ape-man. Another spate of sightings in the region, this time on water-locked Staten Island, between late 1974 and early 1975, involved reports of a huge, hairy creature walking on two legs on roadsides and in the woods. What are we to make of these cases? Even though there remain wooded areas in such urban areas, how could Bigfoot have survived in such populated environments? There is a temptation to delete these reports as they seem to defy logic. We record these cases without further comment and let the reader decide their significance. See, Moravec, Mark, *The UFO-Anthropoid Catalogue: Cases Linking Unidentified Flying Objects and Giant Anthropoid Creatures.* Published by the Australian Centre for UFO Studies, November 1980. Moravec is a psychologist living in Ballarat, Australia.

97. Report #12709 submitted to the Bigfoot Research Organization on

October 4, 2005, and a follow-up interview by BFRO investigator Paul J. Mateja.

98. Report #6700 submitted to the Bigfoot Research Organization on August 4, 2003, and a follow-up interview by an unnamed BFRO investigator.

99. Bigfoot Research Organization report #3399, reported by the witness on November 19, 2001.

100. Report #12709 submitted to the Bigfoot Research Organization on October 4, 2005, and a follow-up interview by BFRO investigator Paul J. Mateja.

101. Interview between Paul Bartholomew and Larry Paap, June 2003.

102. Ripley, Patrick (2003). "Possible Bigfoot Sighting Reported." *The Whitehall Times,* June 12, p. 10.

103. Interview between Paul Bartholomew and Larry Paap, June 2003; Ripley, Patrick (2003). "Possible Bigfoot Sighting Reported." op cit.

104. Interview between the witnesses and Paul Bartholomew.

105. Bigfoot Research Organization report #8905, submitted by the witness on June 24, 2004.

106. Report #13288 submitted to the Bigfoot Research Organization on December 16, 2005, and a follow-up interview by BFRO investigator Paul J. Mateja.

107. Based on a report filed by the Bigfoot Research Organization #15191, submitted by the witnesses on July 13, 2006, and a phone interview by BFRO investigator Paul Kotch.

108. Report filed on September 9, 2006 on the Whitehall, New York Topix site, by Rich Martin, accessed April 15, 2007 at: http://www.topix.net/forum/city/whitehall-ny/TAHSJ7EA1JKVV9IC1.

109. Interview between the witnesses and Paul Bartholomew, September 2006.

110. Interview between Paul Bartholomew and the seventeen-year-old girl, October 2006.

111. Robert, Kenneth Lewis (1937). *Northwest Passage.* Garden City, New York: Doubleday, Doran & Company.

112. Rayno, Paul (1977). "Pioneers and Patriots: More Hairy Beasts." *The Post–Star* (Glens Falls, New York), September 7, p 2.

113. Rayno, Paul, "Pioneers and Patriots—He Lost Old Slippryskin—and Election." March 26, 1975, p. 10, in the *The Post–Star* (Glens Falls, New York).

114. "Slipperyskin—Bear, Bigfoot, or Indian?" *Vermont's Northland Journal,* accessed December 23, 2006 at: http://www.northland journal.com/stories18.html; Rayno (1975). "He Lost Old Slippryskin—and Election." op cit.

115 "Slipperyskin—Bear, Bigfoot, or Indian?" op cit.

116. Rayno (1975). "He Lost Old Slippryskin—and Election." op cit.

117. "Slipperyskin—Bear, Bigfoot, or Indian?" op cit.; Rayno (1975). "He Lost Old Slippryskin—and Election." op cit.

118. Letter from Michael Pluta to Robert Bartholomew, March 2007.

119. *The Commercial Times* (Oswego, New York), August 29, 1861.

120. "Built Like a Horse. Wild Man Creates Terror Among Farmers Around Injun Meadow." *The North Adams Daily Transcript,* August 23, 1895.

121. *Winsted–Lester Prairie Journal,* November 3, 1997.

122. "Wildman Craze of 1895 Made Things a Little Hairy." *Waterbury Republican American* (Connecticut), July 13, 2002.

123. *Winsted–Lester Prairie Journal,* November 3, 1997.

124. "Wild Man's Identity." *The North Adams Transcript,* August 30, 1895, p. 1.

125. "Wild Man's Identity." op cit., p. 1.

126. "Wild Man's Identity." op cit., p. 1.

127. "Wild Man's Identity." op cit., p. 1.

128. "Wild Man Hunt in Haverhill." *Fitchburg Daily Sentinel,* July 14, 1909, p. 7.

129. "Pontoosuc Lake 'Wild Man' Caught After Scuffle with State Patrolman." *The Berkshire Eagle* (Pittsfield, Massachusetts), November 18, 1942, p. 1.

130. Interview between Dr. Warren L. Cook and John Rowell.

131. Story told to John Rowell by Bill Lyford. Rowell, in turn, repeated it to Dr. Cook. Mr. Lyford is now deceased, thus a follow-up interview is not possible.

132. Transcript of a taped interview between Dr. Cook and John Rose of Castleton Corners, Vt., August 1983.

133. Field report #847 filed by The Bigfoot Field Researchers Organization and submitted by the witness on December 2, 2000. BFRO investigator Kevin Withers conducted a follow-up investigation.

134. Sighting report record #296 from the files of the Sasquatch Information Society based in Seattle, Washington, accessed December 27, 2006 at: http://www.bigfootinfo.org/data/bfst.php?srcText_296.

135. Cook, Warren L. (1973). *Flood Tide of Empire: Spain and the Pacific Northwest, 1543–1819.* New Haven, Connecticut: Yale University Press.

136. *Track Record,* newsletter of the Western Bigfoot Society (issue #77, May 1998). This organization is now the International Bigfoot Society (IBF). According to the IBF sighting database, the investigator was Bill Brann.

137. Green (1978). op cit, p. 231.

138. Field report #6496 filed by The Bigfoot Field Researchers Organization, based on a submission sent to the group by the witness, and a follow-up interview by Dr. Wolf H. Fahrenbach of the BFRO.

139. File from the Gulf Coast Bigfoot Research Organization submitted on November 26, 1998. Report taken and processed by Bobby Hamilton. See http://www.gcbro.com/ Vtcal002.htm.

140. Migliore, Mary (1976). "'Bigfoot' Eludes Team on Overnight Campout." *Morning Union,* December 31, 1976.

141. "Bigfoot Story a Joke, Teen is Sorry About Stepping on the Law." United Press International report appearing in the *Kenosha News* (Wisconsin), January 6, 1977.

142. Interview between Dr. Warren Cook and the couple involved.

143. Caraganis, Nick (1977). "Dracut Man Certain he saw Tall, Hairy Creature." *The Sun* (Lowell, Massachusetts), May 17, 1977, p. 3.

144. Caraganis, Nick (1977). op cit.

145. "Lowell Man Flees Hollis After Sighting 'Monster.'" *The Nashua Telegraph,* May 10, 1977.

146. Caraganis, Nick (1977). op cit.

147. Warren Cook learned of this incident by neighbor David Mason. Cook then talked with James Ingalls via the phone.

148. Field report #6643 filed by The Bigfoot Field Researchers Organization, based on a submission sent to the group by the witness, and a follow-up interview by Dr. Wolf H. Fahrenbach of the BFRO. The male witness was about eight years old at the time, and reported the incident in 2003. He said that he and his mother have discussed the episode several times since.

149. Field report #6643. op cit.

150. Transcription by Paul Bartholomew of the second "Sasquatch, alias Bigfoot Conference" held at Castleton State College, Castleton, Vermont, Monday, November 25, 1985, at 8:00 p.m.

151. Duffy, Kevin, "'Goonyak' Reports Debunked, But Rumors of Its Exploits Captivate Northern Vermont," *Rutland Herald,* November 30, 1978.

152. Muscata, Ross A. (2005). "Tales from the Swamp." *The Boston Globe,* October 30, 2005, accessed April 4, 2007 at: http://members.aol.com/soccorro64/ globearticle.htm.

153. Hayward, Ed (1998). "The Bigfoot of Bridgewater; Is it a Man-Beast or Hockomock Crock?" *The Boston Herald,* April 6, accessed April 14, 2007 at: http://www.bigfootencounters.com/articles/bridgewater.htm.

154. Interview between Dr. Cook and the anonymous Fair Haven man.

155. Muscata, Ross A. (2005). "Tales from the Swamp." op cit.

156. Letter from the male witness to Dr. Warren Cook.

157. "Weird Mountain 'Creature' is Reported by Picnickers." *The Berkshire Eagle,* August 23, 1983.

158. Phone interview between Dr. Cook and James Guyette, and several months later in person by Collamer Abbott.

159. Paul Bartholomew transcript of the November 25, 1985, second "Sasquatch, alias Bigfoot Conference" at Castleton State College.

160. Interview by Dr. Cook with Bruce Bateau.

161. Field report #1178 filed by The Bigfoot Field Researchers Organization, based on a submission sent to the group by the adult female witness.

162. Interview between Dr. Cook and Hugo Meyer.

163. Paul Bartholomew transcript of the November 25, 1985, second "Sasquatch, alias Bigfoot Conference" at Castleton State College.

164. Interview between John Rowell and Warren Cook. Rowell was told

the story by a man from Wells River (a part of Newberry, Vermont) who heard it from a man who heard it from the witness.

165. Personal interview by Dr. Cook and several of the witnesses at the scene, Monday, September 23, 1985. Also, transcript of an interview at the location between Paul Bartholomew and Frank "Fron" Grabowski Jr., Ed and Al Davis. Also, Paul Bartholomew's final report on the case.

166. Transcript of interview by Paul Bartholomew and Frank "Fron" Grabowski Jr., Ed and Al Davis, September 23, 1985.

167. "'Big Foot' Seen In West Rutland," *Rutland Herald,* Wednesday, September 25, 1985, p. 20. Paul Bartholomew later talked to a man who claimed that the incident was a hoax, but could offer no substantive evidence.

168. Paul Bartholomew transcript of the November 25, 1985, second "Sasquatch, alias Bigfoot Conference" at Castleton State College.

169. Interview with one of the witnesses from Dr. Warren Cook's personal files.

170. Dr. Cook's personal file. Cook was informed of the sighting by Joe Antell (his student) who was told of it by his Adams Hall suitemate Steven O'Conner, a personal friend of the witnesses.

171. Dr. Cook's interview with his daughter, Susan Cook.

172. Dr. Cook's personal interview with both witnesses. Neither of the freshman girls were aware of Dr. Cook's interest in the phenomena at the time of the sighting.

173. Report #3044 filed for the International Bigfoot Society, May 1998.

174. French, Scott (1987). "The Man Who Spied Bigfoot Comes Forward." *Concord Monitor,* November 13.

175. Interview between one of the teens and Paul Bartholomew.

176. Field report #1180 filed by The Bigfoot Field Researchers Organization, based on a submission to the group by Guy Primo on December 6, 1997.

177. Sighting report record #296 from the files of the Sasquatch Information Society based in Seattle, Washington, accessed December 27, 2006 at: http://www.bigfootinfo.org/data/bfst.php?srcText_296. This was the second sighting by the same witness, the previous one reported in 1969.

178. Report taken by investigator Bobbie Short on November 22, 2000. Accessed on April 14, 2007 at http://www.bigfootencounters.com/sbs/rutland.html.

179. Citro, Joseph (2006). "Population: Is Bigfoot Among Us?" *Livin' The Vermont Way,* September/October 2006: 6–10, 32–33. See p. 33.

180. Report submitted in confidence to the Gulf Coast Bigfoot Research Organization and processed/submitted by Irma Fay Easley, accessed at: http://www.gcbro.com/ NHhills0001.html.

181. Press report on the incident.

182. Report given by the witness to the Sasquatch Information Society based in Seattle, Washington. Report # 331.

183. Report file by witness J. Riggs on December 29, 2006 to the Bigfoot Encounters website operated by Bobbie Short, accessed April 14, 2007 at: http://www.bigfootencounters.com/sbs/orleans.htm.

184. Report submitted in confidence to the Gulf Coast Bigfoot Research Organization and processed/submitted by Irma Fay Easley, accessed at: http://www.gcbro.com/NHcoos 0001.html.

185. Field report #13285 filed by The Bigfoot Field Researchers Organization, based on a submission to the group by the adult male witness on December 15, 2005.

186. Report given by the witness to the Sasquatch Information Society based in Seattle, Washington. Report # 335.

187. Report given by the witness to the Sasquatch Information Society based in Seattle, Washington. Report # 451.

188. Tripp, Rhoda Thomas (compiler). *The International Thesaurus of Quotations.* New York: Thomas Y. Crowell, p. 445.

189. Wilson, Robin (2006). "An Anatomy Professor Tracks Bigfoot." *Chronicle of Higher Education* 52(48): 55.

190. Raeburn, Paul (1990). "Unknown Species of Lion-Headed Monkey Discovered." *The Post–Star* (Glens Falls, New York), June 21, 1990, p. D1.

191. "Rare Tree Expected to have Garden Appeal." *The Southland Times* (New Zealand), February 13, 2001, p. 11. See also: McLoughlin, Stephen, and Vajda, Vivi (2005). "Ancient Wollemi Pines Resurgent." *American Scientist* 93(6):540–547.

192. Jones, Trevor; Carolyn Ehardt, Thomas Butynski, Tim Davenport, Noah Mpunga, Sophy Machaga and Daniela De Luca (2005). "The Highland Mangabey: Lophocebus kipunji: A New Species of African Monkey." *Science 308*(5725): 1161–1164.

193. "70 New Species Discovered in 'Lost World.'" *Geographical* 78(4):6;

McGhee, Karen (2006). "Lost World Brought to Light." *Australian Geographic* 83:22–23; "'Lost World' Found," *Scholastic News*, March 20, 2006:4–5.

194. Lehman, Don (2007). "Police Trace the Murky Trail of a Mountain Man." *The Post–Star* (Glens Falls, New York), January 21.

195. Lehman, Don (2007). op cit.

196. Noory, George (presenter). (2002). Interview on the nationally broadcast United States radio program, "Coast To Coast AM," August 27.

197. Pitcher, John (2001). "Close Encounters." op cit., pp. 1C, 10C.

198. Paul Bartholomew interview with Dr. Cook on May 9, 1989.

199. Letter from Colin Groves to Robert Bartholomew, March 15, 2007.

200. Letter from Colin Groves to Robert Bartholomew, March 15, 2007.

201. Pettifor, Eric (2000). "From the Teeth of the Dragon: Gigantopithecus blacki." In *Selected Readings in Physical Anthropology*. Pp. 143–149. (Peggy Scully, editor). Kendall/Hunt Publishing.

202. Krantz, Grover (1992). op cit.

203. Rattini, Kristin Baird (2005). "Does Bigfoot Exist?" *National Geographic Kids* 349:24–25 (April).

204. Rattini, Kristin Baird (2005). op cit.

205. (2005). "Is it Real? Bigfoot." National Geographic Channel, Night Incorporated. Produced for National Geographic Television and Film.

206. Loxton, Daniel (2005). op cit., p. 103.

207. (Marshall, Laura, producer). "Abominable Snowman: The Search for the Truth." Narrated by Adam Harrington. Learning Channel documentary, circa 1994.

208. "Unsolved Mysteries," NBC Television, airing February 15, 1989.

209. Loxton, Daniel (2005). "Bigfoot Part Two: The Case for the Sasquatch." *The Skeptic* 11(3):96–105. See p. 105.

210. Loxton, Daniel (2005). op cit., p. 105.

211. Krantz, Grover (1992). op cit., p. 125.

212. Loxton, Daniel (2005). op cit., p. 96

213. Loxton, Daniel (2005). op cit., p. 96. Also see: Anonymous (2002). "Grover S. Krantz." *The Times* (London), April 23, p. 34.

214. Krantz, Grover (1992). *Big Foot-Prints: A Scientific Inquiry into the Reality of Sasquatch*. Boulder, Colorado: Johnson Books.

215a. Napier, John (1973). Bigfoot: *The Sasquatch and Yeti in Myth and Reality*. New York: E.P. Dutton, p. 86.

215b. Personal communication from Dr. Janis to Robert Bartholomew, June 10, 2007.

216. Vaughan, Alan, and Peter Guttilla (1998). "A Testable Theory of UFOs, ESP, Aliens, and Bigfoot." *Mutual UFO Network Journal* (November):8–10.

217. Vaughan and Guttilla (1998). op cit.

218. Letter from Bruce Hallenbeck to Robert Bartholomew, April 27, 2007.

219. Loftus, Elizabeth F. (1974). "Reconstructing memory: the incredible eyewitness." *Psychology Today* 8, pp. 116–119; Loftus, E.F. (1979). *Eyewitness Testimony*. Cambridge, MA: Harvard University Press; Ross, D.F., J.D. Read and M.P. Toglia (1994). *Adult Eyewitness Testimony: Current Trends and Developments*. Cambridge: Cambridge University Press.

220. Buckhout, Robert. (1974). Eyewitness testimony. *Scientific American* 231, pp. 23–33; Alcock, J.E. (1981). *Parapsychology: Science or Magic?* New York: Pergamon Press, pp. 73–74; Wells, G. and J. Turtle (1986). "Eyewitness Identification: The Importance of Lineup Models." *Psychological Bulletin* 99: 320–329.

221. Klass, P. (1976). *UFOs Explained*. NY: Random House, pp. 14–15.

222. Bullard, Thomas E. (1982). *Mysteries in the Eye of the Beholder: UFOs and Their Correlates as a Folkloric Theme Past and Present*. Bloomington, Indiana: doctoral dissertation, pp. 10–11.

223. Massad, C.M., M. Hubbard and D. Newston (1979). Selective perception of events. *Journal of Experimental Social Psychology* 15:513–532.

224. Stiles, Fred Tracy (1984). *Old Day—Old Ways: More History and Tales of the Adirondack Foothills*. Hudson Falls, New York: Washington County Historical Society.

225. Shakespeare, William, "A Midsummer Night's Dream." Act 5, scene 1, lines 21–22.

226. Woolridge, A.B. (1987). "The Yeti: A Rock After All." *Cryptozoology* 6:135.

227 Wigen, V. Rae (2005). "An Argument Against Bigfoot." *Skeptic* 12(1):27.

228 Cropper, Paul, and Tony Healy (2006). *The Yowie: In Search of Australia's Bigfoot.* Sydney, Australia: Strange Nation, pp. 185–186; personal communication between Paul Cropper and Robert Bartholomew, April 25, 2007.

229. See Westrum, Ron (1978). "Science and Social Intelligence about Anomalies: The Case for Meteorites." *Social Studies of Science* 8 (4): 461–493.

230. Dendle, Peter (2006). "Cryptozoology in the Medieval and Modern Worlds." *Folklore* 117(August):190–206, see p. 200.

231. Dendle, Peter (2006). op cit.

232. Pitcher, John (2001). op cit.

Bibliography

Alcock, James. 1981. *Parapsychology: Science or Magic?* Pergamon Press, New York.

Anson, Jay. 1977. *The Amityville Horror.* Bantam, New York.

Bartholomew, Paul and Bob, Bruce Hallenbeck and Bill Brann. 1991. *Monsters of the Northwoods.* The authors, Rutland, Vermont.

Biggar, Henry Percival. 1971. *The Works of Samuel de Champlain.* Reprinted, translated and annotated by six Canadian scholars under the general editorship of H.P. Biggar. Volume 1 (1599–1607), translated and edited by H.H. Langton and W.F. Ganong, with the French texts collated by J. Home Cameron. Reprinted by the University of Toronto Press, Toronto, Canada.

Bord, Janet and Colin Bord. 1982. *The Bigfoot Casebook.* Stackpole, Harrisburg, Pennsylvania.

Bruchac, Joseph. 1979. *Stone Giants & Flying Heads.* Adventure Stories of the Iroquois. Crossing Press, Trumansburg, NY.

Bullard, Thomas. 1982. *Mysteries in the Eye of the Beholder: UFOs and Their Correlates as a Folkloric Theme Past and Present* (doctoral dissertation). Indiana University Department of Folklore, Bloomington, Indiana.

Clark, Jerome and Loren Coleman. 1978. *Creatures of the Outer Edge.* Warner, New York.

Cook, Warren.1973. *Flood Tide of Empire: Spain and the Pacific Northwest,* 1543–1819. Yale University Press, New Haven, Connecticut.

Cropper, Paul and Tony Healy. 2006. *The Yowie: In Search of Australia's Bigfoot.* Strange Nation, Sydney, Australia.

Emerson, Ellen. 1884. *Indian Myths Illustrated or Legends, Traditions, and Symbols of the Aborigines of America.* James R. Osgood and Company, Boston, Massachusetts.

Krantz, Grover. 1992. *Big Foot-Prints: A Scientific Inquiry into the Reality of Sasquatch.* Johnson Books, Boulder, Colorado.

Klass, Philip Julian. 1976. *UFOs—Explained.* Random House, New York.

Loftus, Elizabeth. 1979. *Eyewitness Testimony.* Harvard University Press, Cambridge, Massachusetts.

Moravec, Mark. 1980. *The UFO–Anthropoid Catalogue: Cases Linking Unidentified Flying Objects and Giant Anthropoid Creatures.* Published by the Australian Centre for UFO Studies, Gosford, New South Wales, Australia.

Napier, John. 1973. *Bigfoot: The Sasquatch and Yeti in Myth and Reality.* E.P. Dutton, New York.

Pettifor, Eric. 2000. "From the Teeth of the Dragon: Gigantopithecus blacki." In *Selected Readings in Physical Anthropology,* pp. 143–149. (Peggy Scully, editor). Kendall/Hunt Publishing, Dubuque, Iowa.

Roberts, Kenneth Lewis. 1937. *Northwest Passage.* Doubleday, Doran & Company, Garden City, New York.

Ross, David Frank, J. Don Read and Michael P. Toglia. 1994. *Adult Eyewitness Testimony: Current Trends and Developments.* Cambridge University Press, Cambridge, Massachusetts.

Stiles, Fred Tracy. 1984. *Old Day–Old Ways: More History and Tales of the Adirondack Foothills.* Washington County Historical Society, Hudson Falls, New York.

Tripp, Rhoda Thomas (compiler). *The International Thesaurus of Quotations.* Thomas Y. Crowell, New York.

Photograph Credits/Copyrights

Page	Description	Copyright/Credit
2	Sketch of Bigfoot	E. Miner
16	Skenesborough Museum	P. Bartholomew
16	Bigfoot Carving	P. Bartholomew
33	Clifford Sparks	P. Bartholomew
34	Sketch of Bigfoot	R. Dubois
34	Skene Valley sign	P. Bartholomew
36	Sketch of Bigfoot	E. Miner
37	Sketch of two Bigfoot	S. Ellis
40	Sketch of Bigfoot	B. Knights
41	Martha Hallenbeck	P. Bartholomew
41	Bruce Hallenbeck/Doug Hajicek	P. Bartholomew
44	Danny Gordon	P. Bartholomew
48	Possible Footprint	P. Bartholomew
51	Sketch of Bigfoot head	E. Minor
71	Ted Pratt/Dr. Warren Cook	P. Bartholomew
74	Cast	P. Bartholomew
75	Paul Bartholomew at display	P. Bartholomew
86	Dr. Grover Krantz	MH Photo Lib.
87	Cast	P. Bartholomew
88	Ye Ren (Yeren) hair display	P. Bartholomew
100	Paul Cropper/Robert Bartholomew	P. Bartholomew

About the Authors & Editor

Robert Bartholomew holds a Masters Degree in sociology from The State University of New York at Albany, and a Ph.D. in sociology from James Cook University in Queensland, Australia. He is an authority on mass psychology: rumors, panics, scares, hysterias and social delusions. Dr. Bartholomew's articles have appeared in more than fifty scholarly publications including the *British Medical Journal, British Journal of Psychiatry, Canadian Medical Association Journal, Medical Journal of Australia, The International Journal of Social Psychiatry*, and *Psychological Medicine.*

A former university instructor and researcher, he has conducted fieldwork among the Malay people in Malaysia and taught Aborigines in outback Australia. Dr. Bartholomew was a journalist for several New York State radio stations, including WGY, Schenectady, one of the largest stations in the United States. He has been interviewed for articles that were featured in *The Wall Street Journal, The Los Angeles Times, The Chicago Tribune, The San Francisco Chronicle,* and *The New Yorker.* He has also been interviewed on the History Channel.

A licensed high school teacher, his book *Hoaxes, Myths & Manias: Why We Need Critical Thinking* (2003) was endorsed by Robert J. Sternberg, then President of the American Psychological Association. Dr. Bartholomew is the co-author (with Hilary Evans) of *Panic Attacks: Media Manipulation and Mass Delusion* (2004) which Steven Poole of *The Guardian* calls "entertaining...detailed and informative." His book *Exotic Deviance* is a groundbreaking study on the misclassification of mental disorders in non-Western cultures. Professor Arthur Kleinman, chairman of the Department of Social Medicine at Harvard University, calls it "clear, competent and comprehensive."

Paul Bartholomew has been investigating Bigfoot sightings for over thirty years across New York and Vermont. He holds a bachelors degree in journalism from Castleton State College in Castleton, Vermont, where he studied under anthropologist and Pulitzer Prize nominee Dr. Warren L. Cook. Paul worked closely with Professor Cook for several years, investigating regional Bigfoot sightings until his death in 1989, at which time Dr. Cook willed his Bigfoot materials to him in hopes that his work would continue. Paul was a research consultant for the December 2005 History Channel documentary, *Giganto: The Real King Kong* (Whitewolf Entertainment Inc., and in May 2004 for the Outdoor Life Network documentary, *The Creature of Whitehall* (Whitewolf Entertainment Inc.). Paul is a regular guest with popular mystery writer Brad Steiger on the Jeff Rense Radio Show. He has also appeared on The X-Zone Radio Show. For two decades he has spoken on the topic of Bigfoot to museums, schools and civic organizations in New York and Vermont. He has frequently appeared on regional media outlets (television, radio and newspapers). In 2005, Paul made national headlines by successfully lobbying the Whitehall, New York town board to pass an ordinance making it an offense to harm Bigfoot within the township.

Christopher L. Murphy retired from the British Columbia Telephone Company (now Telus) in 1994. He served with this company for 36 years in various management positions including Purchasing Manager and Vendor Quality Manager. He authored four books on business processes and taught a night school course on vendor quality at the British Columbia Institute of Technology. An avid philatelist, Chris has also authored several books on philately.

About a year before he retired, he met the noted sasquatch researcher René Dahinden, and the two became close friends. Chris subsequently republished Roger Patterson's book, *Do Abominable Snowmen of America Really Exist?*, and Fred Beck's book, *I Fought the Apemen of Mt. St. Helens.* Both books were published through Chris' own publishing company, Pyramid Publications (New Westminster, British Columbia).

Chris went on to author, in association with Joedy Cook and George Clappison, *Bigfoot in Ohio: Encounters with the Grassman* (Pyramid Publications, 1997).

In 2000, Chris embarked on a project to assemble a comprehensive pictorial presentation on the sasquatch. This initiative led to a sasquatch exhibit in 2004 at the Vancouver Museum, British Columbia. Chris authored, in association with John Green and Thomas Steenburg, his classic work *Meet the Sasquatch* (Hancock

House Publishers, 2004) to accompany his exhibit. This work received the Anomalist Book of the Year award for that year in the category, Best Illustrated Book.

Chris then wrote a supplement to Roger Patterson's book and it was republished under the title *The Bigfoot Film Controversy* (Hancock House Publishers, 2005). In 2006, Chris updated his Ohio book with his same associates, and it too was published by Hancock House under the title: *Bigfoot Encounters in Ohio: Quest for the Grassman.*

After the Vancouver Museum exhibit ended in February 2005, Chris' exhibit travelled to the Museum of Mysteries in Seattle, Washington, where it was displayed for about five months. Then in 2006, Chris was asked to provide his exhibit to the Museum of Natural History, Pocatello, Idaho (Idaho State University). The exhibit opened on June 16, 2006 and ran until September 3, 2007.

Throughout his 14 years in sasquatch/bigfoot research, Chris has provided many presentations at conferences and has appeared in several television documentaries on this subject. One of his highly noted accomplishments was the construction of a scale model of the Patterson/Gimlin film site. The model was featured in *Fate* magazine (March 2003) in his article "Bigfoot Film Site Insights," and it continues to be a unique attraction at his museum exhibits.

Index

F

G

H

I

J

K

L

M

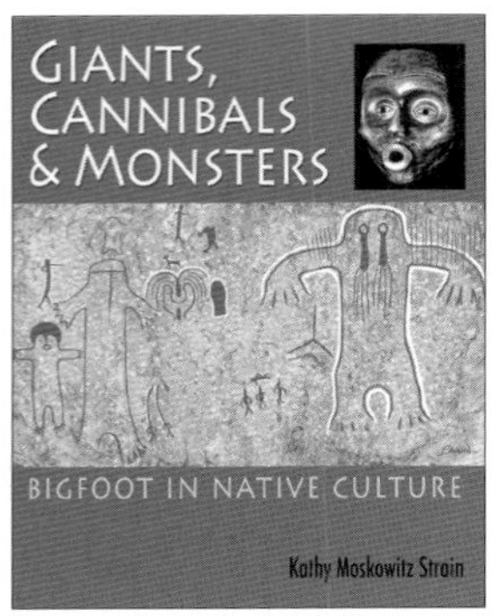

Giants, Cannibals & Monsters
Kathy Moskowitz Strain
978-0-88839-650-1
8½ x 11, sc, 288 pages

The Hoopa Project
Bigfoot Encounters in California
David Paulides
978-0-88839-653-2
5½ x 8½, sc, 336 pages

Know the Sasquatch/Bigfoot
Christopher L. Murphy
978-0-88839-657-0
8½ x 11, sc, 320 pages

Bigfoot Encounters in Ohio
C. Murphy, J. Cook, G. Clappison
0-88839-607-4
5½ x 8½, sc, 152 pages

The Bigfoot Film Controversy
Roger Patterson, Christopher Murphy
0-88839-581-7
5½ x 8½, sc, 240 pages

Sasquatch Bigfoot
Thomas Steenburg
0-88839-312-1
5½ x 8½, sc, 128 pages

In Search of Giants
Thomas Steenburg
0-88839-446-2
5½ x 8½, sc, 256 pages

The Locals
Thom Powell
0-88839-552-3
5½ x 8½, sc, 272 pages

Raincoast Sasquatch
J. Robert Alley
978-0-88839-508-5
5½ x 8½, sc, 360 pages

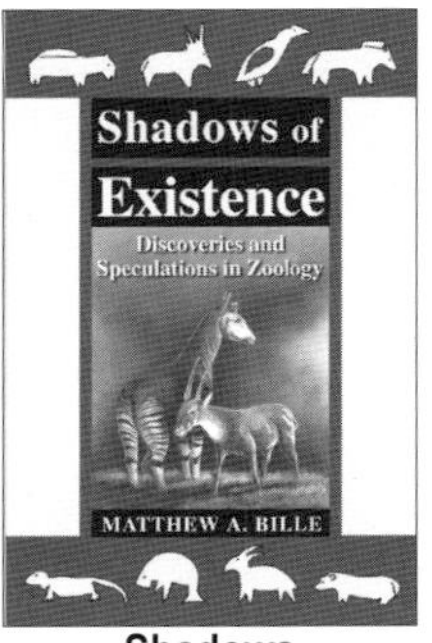

Shadows of Existence
Matthew A. Bille
0-88839-612-0
5½ x 8½, sc, 320 pages

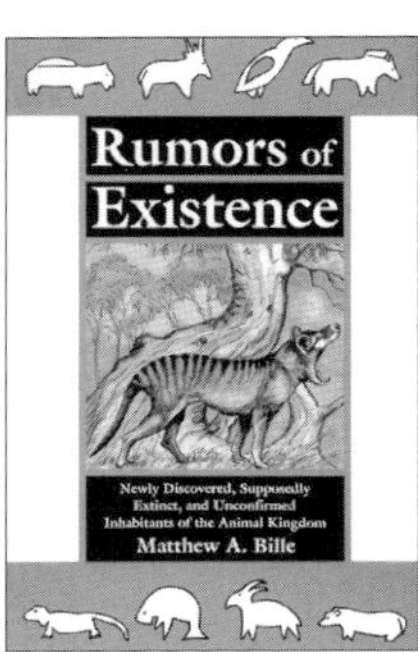

Rumours of Existence
Matthew A. Bille
0-88839-335-0
5½ x 8½, sc, 192 pages

View all **Hancock House** *titles at* **www.hancockhouse.com**